A Text Book Of

SATELLITE COMMUNICATION

(ECE 609)
(ELECTIVE)

Semester - VI

THIRD YEAR DIPLOMA IN ELECTRONICS ENGINEERING GROUPS

As Per SBTE's New Revised Syllabus, Jharkhand

VIJAY G. YANGALWAR

M.Sc., LL.B., L.M.I.S.T.E.
Retd. HOD, Deptt. of Electronics Engg.
Government Polytechnic,
NAGPUR.

NIRALI PRAKASHAN
ADVANCEMENT OF KNOWLEDGE

N4749

Satellite Communication (Sem. VI)   **ISBN 978-93-89944-22-8**

First Edition : February 2020

© : Author

Published By :
NIRALI PRAKASHAN
Abhyudaya Pragati, 1312, Shivaji Nagar
Off J.M. Road, PUNE – 411005
Tel - (020) 25512336/37/39, Fax - (020) 25511379
Email : niralipune@pragationline.com

➢ ## DISTRIBUTION CENTRES

PUNE

Nirali Prakashan : 119, Budhwar Peth, Jogeshwari Mandir Lane, Pune 411002, Maharashtra

(For orders within Pune)   Tel : (020) 2445 2044, Fax : (020) 2445 1538; Mobile : 9657703145

Email : bookorder@pragationline.com, niralilocal@pragationline.com

Nirali Prakashan : S. No. 28/27, Dhyari, Near Pari Company, Pune 411041

(For orders outside Pune)   Tel : (020) 24690204 Fax : (020) 24690316; Mobile : 9657703143

Email : dhyari@pragationline.com, bookorder@pragationline.com

MUMBAI

Nirali Prakashan : 385, S.V.P. Road, Rasdhara Co-op. Hsg. Society Ltd.,

Girgaum, Mumbai 400004, Maharashtra; Mobile : 9320129587

Tel : (022) 2385 6339 / 2386 9976, Fax : (022) 2386 9976

Email : niralimumbai@pragationline.com

➢ ## DISTRIBUTION BRANCHES

JALGAON

Nirali Prakashan : 34, V. V. Golani Market, Navi Peth, Jalgaon 425001,

Maharashtra, Tel : (0257) 222 0395, Mob : 94234 91860

Email : niralijalgoan@pragationline.com

KOLHAPUR

Nirali Prakashan : New Mahadvar Road, Kedar Plaza, 1st Floor Opp. IDBI Bank

Kolhapur 416 012, Maharashtra. Mob : 9850046155

Email : niralikolhapur@pragationline.com

NAGPUR

Nirali Prakashan : Above Maratha Mandir, Shop No. 3, First Floor,

Rani Jhanshi Square, Sitabuldi, Nagpur 440012, Maharashtra

Tel : (0712) 254 7129

DELHI

Nirali Prakashan : 4593/15, Basement, Agarwal Lane, Ansari Road, Daryaganj

Near Times of India Building, New Delhi 110002 Mob : 08505972553

Email : niralidelhi@pragationline.com

BENGALURU

Nirali Prakashan : Maitri Ground Floor, Jaya Apartments, No. 99, 6th Cross, 6th Main,

Malleswaram, Bangaluru 560 003, Karnataka

Mob : +91 9449043034

Email : niralibangalore@pragationline.com

niralipune@pragationline.com | www.pragationline.com

Also find us on www.facebook.com/niralibooks

Preface ...

I take an opportunity to present the text book entitled **"Satellite Communication"** to the students of sixth semester for Electronics Engineering Groups. The main objective of this book is to present the subject matter in a most concise, compact, to the point, simple and lucid manner. The book has been written strictly according to the revised syllabus introduced by SBTE from June 2019.

The book has been written constantly keeping in mind the requirements of all the students regarding the latest and changing trend of the board (SBTE) examinations. It is also written in an easy style with full details and illustrations.

Each chapter gives important points, practice questions and questions asked in board examinations. In short, the book is expected to meet the crying needs of Diploma students of Electronics and Communication Engineering Groups because it gives the theoretical and practical knowledge of Satellite Communication.

The author presents a grateful acknowledgement to various sources, drawn upon in the presentations of this text.

I record my deepest appreciation to my family members for their valuable understanding and co-operation during the spare time which was available tooling to academic, household and social commitment, the book was under preparation. Without their support, patience, love and understanding, I could have not completed this text in a short period of time.

It is my privilege to record indeptness to the publishers Shri. Dineshbhai Furia and Shri. Jignesh Furia for their kind interest in entire work. My thanks are also due to Mr. Rahul Thorat, Mrs. Anjali Muley and Shri. Kiran Velankar of Nirali Prakashan, Pune for publishing this book in good and presentable manner within very short time.

Although every care has been taken to check mistakes and misprints, yet it is very difficult to claim perfection. Any undetected and unintentional errors, omissions, suggestions etc. from students, colleagues, teachers and practicizing engineers for improvements brought to are notice in good spirit are most welcome.

Vijay G. Yangalwar

Syllabus ...

1. Introduction to Satellite Communication (Hrs. 08)

Origin, Brief History, Current state and Advantages of Satellite Communication, Active & Passive satellite, Orbital aspects of Satellite Communication, Angle of Elevation, Propagation Delay, Orbital Spacing, System Performance.

2. Satellite Link Design (Hrs. 12)

Link design equation, System noise temperature, C/N & G/T ratio, Atmospheric and Ionospheric effects on link design, Complete link design, Interference effects on complete link design, Earth station parameters. Earth space propagation effects, Frequency window. Free-space loss, Atmospheric absorption, Rainfall attenuation, Ionospheric scintillation, Telemetry, Tracking and command of satellites.

3. Satellite Multiple Access System (Hrs. 12)

FDMA techniques, SCPC and CSSB systems. TDMA frame structure, Burst structure, Frame efficiency, Super-frame, Frame acquisition and synchronization. TDMA versus FDMA, Burst time plan, Beam hopping. Satellite switched, Erlang call congestion formula, DA-FDMA, DA-TDMA.

4. Satellite Services (Hrs. 10)

INTELSAT, INSAT Series, VSAT, Weather forecasting, Remote sensing. LANDSAT, Satellite Navigation, Mobile Satellite Service.

5. Laser and Satellite Communication

Laser & Satellite Communication link analysis, Optical satellite link, Tx & Rx, Satellite, Beam acquisition, Tracking & pointing, Cable channel frequency, Head end equation, Distribution of signal, Network specifications and architecture, Optical fibre CATV system.

☆☆☆

Contents ...

INTRODUCTION TO SATELLITE COMMUNICATION

Origin, Brief History, Current state and Advantages of Satellite Communication, Active & Passive satellite, Orbital aspects of Satellite Communication, Angle of Elevation, Propagation Delay, Orbital Spacing, System Performance.

1.1 SATELLITE COMMUNICATION

1.1.1 Definition

- **If the communication takes place between any two Earth stations through a satellite, then it is called as satellite communication.**

- In this communication, electromagnetic (RF) waves are used as carrier signals. These signals carry the information such as voice, audio, video or any other data between ground and space and vice-versa.

1.1.2 Origin

- The concept of the geostationary communications satellite was first proposed by Arthur C. Clarke, along with *Vahid K. Sanadi* building on work by *Konstantin Tsiolkovsky*.

- In October 1945, *Arthur C. Clarke* published an article titled *"Extraterrestrial Relays"* in the British magazine *'Wireless World'*.

- The article described the fundamentals behind the deployment of artificial satellites in geostationary orbits for the purpose of relaying radio signals. Thus, *Arthur Clarke* is often quoted as being the inventor of the communication satellite.

1.1.3 Brief History

- The first artificial Earth satellite was *Sputnik 1*, which was put into orbit by the *Soviet Union* on October 4, 1957. It was equipped with an on-board radio-transmitter that worked on two frequencies: 20.005 MHz and 40.002 MHz.

- *Sputnik 1* was launched as a major step in the exploration of space and *rocket* development. However, it was not placed in the orbit for the purpose of sending data from one point on Earth to another.

- The first satellite to relay communications was *Pioneer-1*, an intended lunar probe. The first satellite purpose-built to relay communications was NASA's Project SCORE in 1958, which used a *tape recorder* to store and forward voice messages. It was used to send a Christmas greeting to the world from U.S. President *Dwight D. Eisenhower*. *Courier 1B*, built by *Philco*, launched in 1960, was the world's first *active repeater satellite*.

- The first *artificial satellite* used solely to further advances in global communications was a *balloon* named *Echo-1*. *Echo-1* was the world's first artificial communications satellite capable of relaying signals to other points on Earth. It soared 1,600 kilometres (1,000 mi) above the planet after its August 12, 1960 launch.

- The world's first *inflatable satellite* or *"satelloon"*, as they were informally known-helped lay the foundation of today's satellite communications.

- The idea behind a communications satellite is simple: Send data up into space and beam it back down to another spot on the globe.
- Echo-1 accomplished this by essentially serving as an enormous mirror, 10 stories tall, that could be used to reflect communication's signals. On July 10, 1962, NASA launched a first privately sponsored satellite from Cape Canaveral. Relay-1 was launched on December 13, 1962, and it became the first satellite to transmit across the Pacific Ocean on November 22, 1963.
- An immediate antecedent of the *geostationary satellites* was the Hughes Aircraft Company's Syncom-2, launched on July 26, 1963, which was the first communications satellite in a *'geostationary orbit'*. All *Mars landers*, arrived from Patifinders, have used orbiting spacecraft as communications satellites for relaying their data to Earth.

1.1.4 Need

1. The satellites provide communication for long distances (above 1500 km), which is well beyond the line-of-sight.
2. Since the satellites are located at certain height above Earth, the communication takes place between any two Earth stations easily via satellite. So, it overcomes the limitations of communication between two Earth stations due to Earth's curvature.
- Due to above key features, there is a need of satellite communication.

1.1.5 Working Principle

- The basic principle of satellite communication is illustrated in Fig. 1.1. It consists of a communication satellite and two Earth stations.

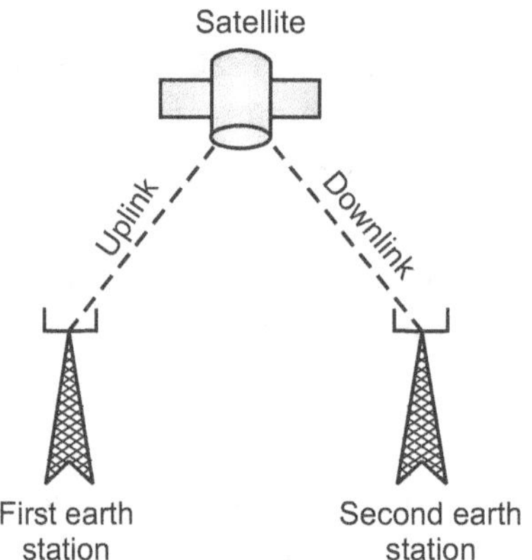

Fig. 1.1 : Satellite Communication

- A repeater is a circuit, which increases the strength of the received signal and then transmits it. But, this repeater works as a *transponder*. That means, it changes the frequency band of the transmitted signal from the received one.
- The frequency with which, the signal is sent into the space is called as *Uplink frequency*. Similarly, the frequency with which, the signal is sent by the transponder is called as *Downlink frequency*.
- The uplink frequency is the frequency at which, the first Earth station is communicating with the satellite. The satellite transponder converts this signal into another frequency and sends it down to the second Earth station.
- This frequency is called as *Downlink frequency*. In similar way, second Earth station can also communicate with the first one.
- The process of satellite communication begins at an Earth station. Here, an installation is designed to transmit and receive signals from a satellite in an orbit around the Earth. Earth stations send the information to satellites in the form of high powered, high frequency (GHz range) signals.

- The satellites receive and transmit the signals back to Earth, where they are received by other Earth stations in the coverage area of the satellite. Satellite's *footprint* is the area which receive a signal of useful strength from the satellite.

1.1.6 Advantages

1. Area of coverage is more than that of terrestrial systems.
2. Each and every corner of the Earth can be covered.
3. Transmission cost is independent of coverage area.
4. More bandwidth and broadcasting possibilities.

1.1.7 Disadvantages

1. Launching of satellites into orbits is a costly process.
2. Propagation delay of satellite systems is more than that of conventional terrestrial systems.
3. Difficult to provide repairing activities, if any problem occurs in a satellite system.
4. Free space loss is more.
5. There can be congestion of frequencies.

1.1.8 Applications

1. Radio broadcasting and voice communications.
2. TV broadcasting, such as Direct To Home (DTH).
3. Internet applications, such as providing Internet connection for data transfer, GPS applications, Internet surfing, etc.
4. Military applications and navigations.
5. Remote sensing applications.
6. Weather condition monitoring and forecasting.

1.2 COMMUNICATION'S SATELLITES

1.2.1 Definition

- **A communication's satellite is an artificial satellite that relays and amplifies radio telecommunications signals via a transponder; which creates a communication channel between a source transmitter and a receiver at different locations on Earth.**
- Communication's satellites are used for television, telephone, radio, internet, and military applications.

1.2.2 Classes

1. Passive satellites
2. Active satellites

1.2.3 Passive Satellites

- Passive satellites only reflect the signal coming from the source, towards the direction of the receiver.
- With passive satellites, the reflected signal is not amplified at the satellite, and only a very small amount of the transmitted energy actually reaches the receiver. Since the satellite is so far above the Earth, the radio signal is attenuated due to free-space path loss, so the signal received on the Earth is very, very weak.
- The communication's satellite, which only reflect the signal coming from the source, towards the direction of the receiver is called a passive satellite.

1.2.4 Active Satellites

- **The communication's satellite which amplify the received signal before transmitting it to the receiver on Earth is called an active satellite.**
- Active satellites amplify the received signal before transmitting it to the receiver on Earth. Telstar was the second active, direct relay communication's satellite.

1.3 SATELLITE ORBITS

1.3.1 Definition

- **The path of satellite revolving around the Earth is known as an orbit.**
- The path can be represented with mathematical notations.
- **The orbits, which are assigned to satellites with respect to Earth are called as Earth's orbits.**

1.3.2 Orbital Aspect

- We know that satellite revolves around the Earth, which is similar to the Earth revolves around the Sun. So, the principles which are applied to Earth and its movement around the Sun are also applicable to satellite and its movement around the Earth. *Kepler's laws* are helpful to visualize the motion through space.

  1. The path followed by a satellite around its primary (the Earth) will be an ellipse.
  2. For equal intervals of time, the **area** covered by the satellite will be same with respect to centre of mass of the Earth.
  3. The square of the periodic time of an elliptical orbit is proportional to the cube of its semi-major axis length.

- A communication's satellite has to balance the gravitational force and pulling force to keep itself in its orbit.

1.3.3 Look Angle

- **Azimuth angle and elevation angle of the Earth station antenna combined together is called look angle.**
- Generally, the values of these angles change for **non-geostationary orbits**, whereas, the values of these angles do not change for **geostationary orbits**. Because, the satellites present in geostationary orbits appear stationary with respect to Earth.

1. **Azimuth Angle:**

- **The angle between local horizontal plane and the plane passing through Earth station, satellite and center of Earth is called as azimuth angle.** It is given by

$$\alpha = 180° + \tan^{-1}\frac{\tan G}{\tan L} \qquad \qquad ...(1.1)$$

  where, L = Latitude of Earth station antenna.

  G = the difference between position of satellite orbit and Earth station antenna.

2. **Elevation Angle:**

- **The angle between vertical plane and line pointing to satellite is known as Elevation angle.**
- Vertical plane is nothing but the plane, which is perpendicular to horizontal plane. It is given by,

$$\beta = \tan^{-1}\frac{\cos G \cdot \cos L - 0.15}{\sqrt{1 - \cos^2 G \cdot \cos^2 L}} \qquad \qquad ...(1.2)$$

1.3.4 Orbital Slots/Spacings

- **The locations specified in degree of longitude on the geosynchronous orbit are known as orbital slots or spacings.**
- The FCC and ITU have progressively reduced the required spacing down to only 2° for C-band and ku-band satellite due to the huge demand for orbital slots.

1.3.5 Propagation Delay

- **In a communication system, propagation delay refers to the time lag between the departure of a signal from the source and the arrival of the signal at the destination.**

- The propagation delay can range from a few nanoseconds (ns) or microseconds (μs) in local area networks (LANs) upto about 0.25 s in geostationary-satellite communication's systems. Additional propagation delays can occur as a result of the time required for packets to make their way through land-based cables and nodes of the Internet.

1.3.6 System Performance

- **System performance** focuses on quantitative design criteria for embedded applications and their resource utilisation in trade-off with cost. Covering a broad range of extra-functional criteria, system performance may, for example, relate to perceived quality of functional outcomes, timing behaviour, storage space and bandwidth utilization, energy consumption and heat dissipation. Such cross-cutting criteria are among the most important ones for embedded systems as they are often a key selling factor.

- Performance criteria are either expressed as constraints or as optimization objectives. Design for system performance focusses on satisfying (often conflicting) extra-functional requirements and is determined by a complex coordination and balancing of inter-related events in subsystems and components. Achieving the desired system performance involves trade-offs between all disciplines involved in the design.

- The system has the following parameters and factors, which affect the performance of a communication system:

    1. **Redundancy:** More redundancy bits have more error correction capabilities, if an optimized encoding algorithm is used. High redundancy requires high bandwidth so while designing an error correction code, it was focussed that with low redundancy; maximum number of errors must be corrected.

    2. **BER (Bit Error Ratio):** In digital communication, **BER is defined as the ratio of number of bit errors to the total number of transferred bits during a studied time interval.** During the transmission, number of bits in communication channel has been altered due to *noise, interference, distortion* or *bit synchronization errors*. While designing a communication system or error control codes it was focused that BER must be minimized.

    3. **Extensibility:** The communication system or error control code can be expanded, and improved when required and error correction capabilities could be increased.

    4. **Modularity:** The communication system consists of individual blocks in its construction and every module or entity treats the incoming bits or signals from one module to the other as an output source.

    5. **Usability:** The error control code widely designed for space and satellite communication but it can be in mobile communication systems and in other communication networks.

Practice Questions

1. Write a short note on history of satellite communication.

2. State the origin of satellite communication.

3. What is satellite communication?

4. What is the need of satellite communication?

5. State advantages and disadvantages of satellite communication.

6. State applications of satellite communication.

7.   Describe passive and active satellites.

8.   What is an orbit ?

9.   Describe orbital aspect of satellite communication.

10.   Explain all types of orbits.

11.   Explain all orbital elements.

12.   Define the following terms:

     (a) Azimuth angle, (b) Elevation angle, (c) Lock angle

13.   Explain orbital effects in communication system performance.

14.   Explain propagation delay in radio communication.

15.   Write a short note on the following:

     (a) Propagation delay, and (b) Orbital spacing

16.   What is system performance?

17.   List the factors which affect the performance of a system.

☆☆☆

Chapter 2

SATELLITE LINK DESIGN

Syllabus

Link design equation, System noise temperature, C/N & G/T ratio, Atmospheric and Ionospheric effects on link design, Complete link design, Interference effects on complete link design, Earth station parameters. Earth space propagation effects, Frequency window. Free-space loss, Atmospheric absorption, Rainfall attenuation, Ionospheric scintillation, Telemetry, Tracking and command of satellites.

2.1 LINK DESIGN EQUATION

2.1.1 Introduction

- A satellite communication network consists of a number of Earth stations interconnected via a satellite. The radio link used for interconnections are designed so as to deliver messages at the destination with acceptable fidelity.

- A compromise is exercised between the *quality* and *quantity* of delivered messages and practical constraints, such as economics and the state of technology. To deliver a large amount of information at a *very high quality* may require unacceptably high cost. Factors which need consideration in a *link design* include operational frequency, propagation effects, acceptable spacecraft/ground terminal complexity (hence cost), effects of noise and regulatory requirements.

2.1.2 Need

- The communication link between a satellite and the Earth Station (ES) is exposed to a lot of impairments such as noise, rain and atmospheric attenuations.

- It is also prone to loss, such as those resulting from antenna misalignment and polarization. It is, therefore, crucial to design for all possible attenuation scenarios before the satellite is deployed.

- Therefore, there is a need for proper design of a satellite link for communication.

2.1.3 Definition of Satellite Link

- *The satellite link is essentially a radio relay link, much like the terrestrial microwave radio relay link with the singular advantage of not requiring as many retransmitters as are required in the terrestrial link.*

2.1.4 Link Design Equation

- The design of satellite link is quite important as it gives the estimate of power that the satellite would be able to receive from transmitting Earth satellite and power received from satellite repeater by the receiving Earth station.

  The power received by the receiving antenna is given by:

$$P_R = \frac{P_T \cdot A_R}{A_O} \qquad ...(2.1)$$

The directivity of antenna is described by its gain as:

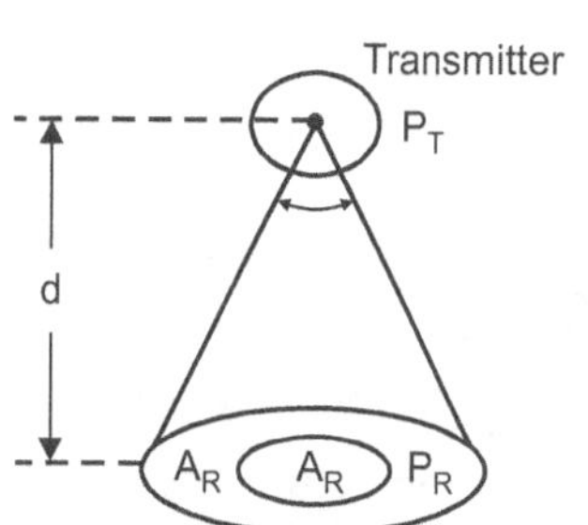

Fig. 2.1 : Satellite link design

$$G_T = \frac{4\pi d^2}{A_O} \qquad ...(2.2)$$

Substituting equation (2.2) in equation (2.1), we get

$$P_R = \frac{P_T \cdot G_T \cdot A_R}{4\pi d^2} \qquad \qquad ...(2.3)$$

The product $P_T G_T$ is called effective isotropic radiated power (EIRP) of the transmitter and is a figure of merit for a transmitter. The receiving antenna gain G_R is related to the effective area A_R by the relationship,

$$G_R = \frac{4\pi A_R}{\lambda^2} \qquad \qquad ...(2.4)$$

Substituting equation (2.4) in equation (2.3), we get,

$$P_R = P_T G_T G_R \left(\frac{\lambda}{4\pi d}\right)^2 \qquad \qquad ...(2.5)$$

$$\therefore \qquad \frac{P_T}{P_R} = \frac{1}{G_T G_R \left[\dfrac{\lambda}{4\pi d}\right]^2} \qquad \qquad ...(2.6)$$

$$\alpha_{dB} = 10 \log_{10}\left(\frac{P_T}{P_R}\right) \qquad \qquad ...(2.7)$$

$$\alpha_{dB} = 22 + 20 \log_{10}\left(\frac{d}{\lambda}\right) - G_T - G_R \qquad \qquad ...(2.8)$$

$$\therefore \qquad L_{FS} = 22 + 20 \log_{10}\left(\frac{d}{\lambda}\right) \text{ dB} \qquad \qquad ...(2.9)$$

where L_{FS} = free space loss

2.2 SYSTEM NOISE TEMPERATURE

2.2.1 Noise Temperature

- **A noise temperature is the temperature of a resistor that has noise power equal to that of the device or circuit.**
- Specifically, the noise temperature T is defined by,

$$T = \frac{P_N}{KB} \qquad \qquad ...(2.10)$$

where,

P_N = Noise power within bandwidth B.

K = Boltzmann's constant = $1.38 \times 10^{-23} \text{J K}^{-1}$

B = Total bandwidth over which the noise power is measured

- Thus, the noise temperature is proportional to the power spectral density $\frac{P_N}{B}$, i.e. the power that would be absorbed from the component or source by a matched load.
- Noise temperature is generally a function of frequency, unlike that of an ideal resistor, which is simply equal to the actual temperature at all frequencies.

2.2.2 Communication System Noise Temperature

- A communication system is typically made up of a *transmitter*, a *communication channel*, and a *receiver*. The communication channel may consist of a combination of different physical media, resulting in an electrical signal presented to the receiver. Whatever physical media a channel consists of, the transmitted signal will be attenuated and corrupted with additive noise.
- The additive noise in a receiving system can be of *thermal origin* or can be from other noise-generating processes. The noise power spectral density generated by all these sources $\left(\frac{P_N}{B}\right)$ can be described by

assigning to the noise a temperature given by equation (2.10). In a wireless communications receiver, the equivalent input noise temperature T_{eq} would be equal to the sum of two noise temperatures.

$$T_{eq} = T_{ant} + T_{sys} \qquad \qquad ...(2.11)$$

- The antenna noise temperature T_{ant} gives the noise power seen at the output of the antenna. The noise temperature of the receiver circuitry T_{sys} represents noise generated by noisy components inside the receiver.

- Note that T_{eq} refers not to the noise at the output of the receiver after amplification, but the equivalent input noise power. In other words, the output of the receiver reflects that of a noiseless amplifier whose input had a noise level not of T_{ant} but of T_{eq}.

- Thus the *figure of merit* of a communications system is not the noise level at the speaker of a radio. The additional level added by the receiver to the original noise level before its gain was applied is given by BKT_{sys}. If a signal is present, then the decrease in signal to noise ratio incurred using the receiver system

with a noise temperature of T_{sys} is proportional to $\dfrac{1}{T_{ant}} - \dfrac{1}{(T_{ant} + T_{sys})}$ $\qquad ...(2.12)$

2.2.3 $\dfrac{C}{N}$ Ratio and $\dfrac{G}{T}$ Ratio

- The antenna gain to noise temperature ratio G/T is a figure of merit to indicate the performance of the Earth station antenna and the low noise amplifier in relation to sensitivity in receiving down link carrier from the satellite.

- The parameter G is the receiver antenna gain referred to the input of the low noise amplifier. The parameter T is defined as the Earth station system noise temperature referred also to the input of the low noise amplifier.

$$\frac{G}{T} = [G_r x] - 10 \log (T_S) \qquad \qquad ...(2.13)$$

where:

$\quad G_r x$ = Receiver gain in dB

$\quad T_S$ = System noise temperature in $^\circ K$

- In the link equation, by unfolding the KTB product under the logarithm, the link equation becomes,

$$\frac{C}{N} = [EIRP] - [Losses] + [G] - 10 \log (K) - 10 \log (T_S) - 10 \log (B) \quad ...(2.14)$$

The difference $[G] - 10 \log (T_S)$, is the figure of merit.

$$\frac{C}{N} = [EIRP] - [Losses] + \frac{G}{T_S} - 10 \log (K) - 10 \log (B) \qquad \qquad ...(2.15)$$

where:

$\quad K$ = Boltzmann's constant

$\quad B$ = Carrier occupied bandwidth

We know, $\qquad \qquad P_N = KTB = N_0 B \qquad \qquad ...(2.16)$

$\therefore \qquad \qquad \dfrac{C}{N} = \dfrac{C}{N_0 B} \, dB \qquad \qquad ...(2.17)$

or $\qquad \qquad \dfrac{C}{N} = \dfrac{C}{N_0} - B \qquad \qquad ...(2.18)$

Here the unit of $\dfrac{C}{N_0}$ is in dB Hz, $\dfrac{C}{N}$ in dB .

Hence, $\qquad \qquad \dfrac{C}{N} = \dfrac{C}{N_0} - 10 \log (360 \times 10^6) \qquad \qquad ...(2.19)$

2.3 ATMOSPHERIC AND IONOSPHERIC EFFECTS ON LINK DESIGN

2.3.1 Atmospheric Effect

- Attenuation by atmospheric gates at microwave and millimetric frequencies is mainly due to oxygen and water vapour absorption. Oxygen possess a permanent *magnetic moment* and because of the interaction of this moment with the magnetic field of the wave *absorption* of wave energy takes place.

- At frequencies below 36 Hz, path attenuation due to atmospheric gases, rain and clouds is small and is often neglected, whereas oxygen and water vapour in the lower atmosphere significantly affect path attenuation at higher frequencies. As the effect is highly frequency dependent, so the attenuation due to atmospheric absorption in some frequency bands is much greater than in others.

- Below 22.3 GHz, the specific attenuation increases with frequency tremendously and it can be more than 10 times higher at 15 GHz than at 2 GHz. Also, the gaseous absorption is less than 1 dB for most paths below 100 GHz.

- Taking into account the relative contribution, it is obvious that there will be more attenuation in the presence of water vapour than in dry air because of the presence of more molecules in water vapour.

2.3.2 Ionospheric Effect

- Ionosphere is the upper part of the atmosphere, where sufficient ionization exists to influence radio wave propagation. The ionosphere usually consists of two layers: the E layer which is about 80 to 113 km above the Earth's surface and reflects radio waves of lower frequency.

- Above E layer is F layer, which reflects higher frequency radio waves. The F layer is then further sub-divided into F1 and F2 layers.

- The F1 layer is lower portion of F layer and exists from 150 to 200 km above the Earth's surface, whereas F2 layer is the upper portion and exists at a height of 200 to 500 km. F2 layer is mainly responsible for reflection of high frequency waves during day and night. Since the *ionization* is mainly caused by solar radiations, it is dependent on location, time of the day, season and *Sunspots*.

- Radio waves propagating through ionosphere experience different *attenuation mechanisms* such as absorption, reflection, refraction, scattering, polarization, group delay and fading/scintillation. In the region other than ionosphere i.e. troposphere, stratosphere etc. radio waves loses its energy mainly due to absorption, cloud and rain *attenuation*, attenuation due to snow, hail and fog. Rain is considered to be the major cause of attenuation at frequencies above 10 GHz.

- As reported in literature, atmosphere contains free electrons, ions, and molecules and their interaction with radio waves depend strongly on frequency, so as the frequency increases, the effect of attenuation also increases.

2.3.3 Rainfall Attenuation

- The strength of satellite signal may be degraded or reduced under rain conditions; in particular radio waves above 10 GHz are subject to attenuation by *molecular absorption* and rain. Presence of rain drops can severely degrade the reliability and performance of communication links.

- Attenuation due to rain effect is a function of various parameters including elevation angle, carrier frequency, height of Earth station, latitude of Earth station and rain fall rate.

- The primary parameters, however, are drop-size distribution and the number of drops that are present in the volume shared by the wave with the rain. It is important to note that, attenuation is determined not by how much rain has fallen, but the rate at which it is falling.

- Rain attenuation is a key limiting factor in using high frequency bands in satellite and terrestrial microwave systems. Rain drops both absorb and scatter radio-wave energy. Very intense rain rate may cause link outage. If the rain drop size approaches half the wavelength of the signal in diameter, the signal will be attenuated. Higher frequencies exhibit more attenuation than lower frequencies due to smaller wavelength.

- Rain degrades the performance of a satellite communication system by increasing the noise temperature of the Earth station antenna. While raining, it receives thermal radiation from rain drops, which cause an increase in the overall noise temperature.

2.3.4 Atmospheric Absorption

- The atmosphere is a layer of gases covering the surface of the Earth. There are many gases in the atmosphere. The two that cause almost all of the radio wave attenuation are oxygen and water vapour. While oxygen is an invariant parameter that can be easily accounted for and does not change with latitude or season, water vapour is highly variant; the diurnal and seasonal variability of water vapour can be extremely large.

- The *electric dipole* associated with the water vapour molecule interacts with the incident electromagnetic radiation to produce rotation absorption lines at 22.2, 183.3 and 324 GHz. The two components due to oxygen and water vapour should be added together for the total specific attenuation.

- The water vapour absorption is considered at Ku and Ka bands, but oxygen absorption is not considered, as *Fixed Satellite Services* (FSS) do not employ the frequencies close to the oxygen absorption band about 60 GHz.

2.3.5 Ionospheric Scintillation

- Ionosphere is located from 100 to 1000 km above the Earth and filled with a large number of electrons and plasmas.

- ***Irregularities in the ionosphere could cause inhomogeneties in the refractivity of RF signals, while the signals are travelling through the ionosphere, which is known as ionospheric scintillation.***

- Ionospheric scintillation is ultimately the function of phase and amplitude caused by irregularities in the Earth's ionosphere.

- Ionospheric scintillation could affect radio signals ranging from 10 MHz to 10 GHz, and the phenomenon varies with time and space. General speaking, *scintillation* is severer in the equatorial area and high-latitude regions than in the middle-latitude area.

- The performance of satellite communication could be degraded due to the presence of ionospheric scintillation. Global Navigation Satellite System (GNSS) satellite-receiver links travelling through the atmosphere are also vulnerable to ionospheric scintillation. This could result in some inevitable harmful effects such as signal fading and loss of signal tracking.

- Signal fading could cause a decline of signal-to-noise ratio (SNR), which could result in degradation of positioning accuracy, and loss of signal tracking could lead to navigation system failure. Moreover, scintillation could also increase the errors of pseudo range and carrier-phase range measurement, and thus have harmful impacts on high precision positioning.

2.3.6 Free-Space Loss

- **In telecommunications, the Free-Space Path Loss (FSPL) is the attenuation of radio energy between the feedpoints of two antennas that results from the combination of the receiving antenna's capture area plus the obstacle free, line-of-sight path through free space usually air.**

- **Standard Definition:** Free-space loss is the loss between two isotropic radiators in free space, expressed as a power ratio.

- Free-space loss does not include any power loss in the antennas themselves due to imperfections such as resistance. Free-space loss increases with the square of distance between the antennas because the radio waves spread out due the inverse square law and decreases with the square of the wavelength of the radio waves.

- The Free-Space Path Loss (FSPL) is rarely used standalone, but rather as a part of gain of antennas. It is a factor that must be included in the power link budget of a radio communication system, to ensure that sufficient radio power reaches the receiver that the transmitted signal is received intelligibly.

- The free-space path loss for isotropic antenna is given by,

$$\text{FSPL} = \left(\frac{4\pi d}{\lambda}\right)^2 = \left(\frac{4\pi df}{c}\right)^2 \qquad \text{...(2.20)}$$

where: f = Frequency of radio waves,

$$\lambda = \text{Wavelength of radio waves} = \frac{f}{c}$$

d = Distance between antennas

2.4 COMPLETE LINK DESIGN

2.4.1 Design Methodology

- The design methodology for a one-way satellite communication link can be summarized into the following steps. The return link follows the same procedure.

  Step 1: Frequency band determination.

  Step 2: Satellite communication parameter determination.

  Step 3: Earth station parameter determination; both uplink and downlink.

  Step 4: Establish uplink budget and a transponder noise power budget to find $(C/N)_{up}$ in the transponder.

  Step 5: Determine transponder output power from its gain or output backoff.

  Step 6: Establish a downlink power and noise budget for the receiving Earth station

  Step 7: Calculate $(C/N)_{down}$ and $(C/N)_{up}$ for a station at the outermost contour of the satellite footprint.

  Step 8: Calculate SNR/BER ratio in the baseband channel.

  Step 9: Determine the link margin.

  Step 10: Do a comparative analysis of the result vis-a-vis the specification requirements.

  Step 11: Tweak system parameters to obtain acceptable $(C/N)_0$, SNR and BER values.

  Step 12: Propagation condition determination.

  Step 13: Uplink and downlink unavailability estimation.

2.4.2 Complete Design Equation

- In satellite communication systems, there are two types of power calculations. There are transmitting power and receiving power calculations. In general, these calculations are called as *Link budget calculations*.

- Assume that an isotropic radiator (antenna) is situated at the center of the sphere having radius, r. We know that power flux density is the ratio of power flow and unit area. Power flux density, Ψ_i of an isotropic radiator is given by,

$$\Psi_i = \frac{P_s}{4\pi r^2} \qquad \text{...(2.21)}$$

where, P_s = power of a transmitter

In general, the power flux density of a practical antenna varies with direction. But, its maximum value will be in one particular direction only.

- The gain of practical antena is defined *as the ratio of maximum power flux density of practical antenna and power flux density of an isotropic antenna*. Therefore, the gain of an antenna or antenna gain, G is given by,

$$G = \frac{\Psi_m}{\Psi_i} \qquad \text{...(2.22)}$$

where, Ψ_m = Maximum power flux density of practical antenna.

Ψ_i = Power flux density of an isotropic radiator (antenna).

- Equivalent isotropic radiated power (EIRP) is the main parameter that is used in the measurement of link budget. Mathematically, it can be written as

$$\text{EIRP} = GP_s \qquad \text{...(2.23)}$$

where, G = Gain of transmitting antenna,

$\quad\quad P_s$ = Power of a transmitter.

- The difference between the power sent at one end and received at the receiving station is known as *transmission loss*. The losses are constant losses and *variable losses*.

- The losses which are constant such as feeder losses are known as *constant losses*. No matter what precautions we might have taken, still these losses are bound to occur.

- Another types of losses are *variable losses*. The sky and weather condition is an example of this type of loss. Means, if the sky is not clear, signal will not reach effectively to the satellite or vice versa.

- Therefore, our procedure includes the calculation of losses due to clear weather or clear sky condition as first step because these losses are constant. They will not change with time. Then in the second step, we can calculate the losses due to foul weather condition.

- Earth station uplink is the process in which Earth is transmitting the signal to the satellite and satellite is receiving it. Its mathematical equation can be written as

$$\left(\frac{C}{N_0}\right)_U = [\text{EIRP}]_U + \left(\frac{G}{T}\right)_U - [\text{Losses}]_U - K \qquad \text{...(2.24)}$$

where,

$\left(\dfrac{C}{N_0}\right)$ = Carrier to noise density ratio

$\left(\dfrac{G}{T}\right)$ = Satellite receiver G/T ratio and units are dB/K

U = Uplink phenomenon.

- Here, losses represent the satellite receiver feeder losses. The losses which depend upon the frequency are all taken into the consideration.

- In satellite downlink process, satellite sends the signal and the Earth station receives it. The equation is same as the satellite uplink with a difference that we use the abbreviation "D" everywhere instead of "U" to denote the downlink phenomena. Its mathematical equation can be written as,

$$\left(\frac{C}{N_0}\right)_D = [\text{EIRP}]_D + \left(\frac{G}{T}\right)_D + [\text{Losses}]_D - K \qquad \text{...(2.25)}$$

Here, all the losses that are present around Earth stations. In equation (2.23), we have not included the signal bandwidth B. However, if we include then the losses will be modified as follows.

$$\left(\frac{C}{N_0}\right)_D = [\text{EIRP}]_D + \left(\frac{G}{T}\right)_D + [\text{Losses}]_D - K - B \qquad \text{...(2.26)}$$

- If we are taking ground satellite into consideration, then the free-space spreading loss (FSP) should also be taken into consideration.

- If antenna is not aligned properly, then losses can occur. So, we take AML (antenna misalignment losses) into account. Similarly, when signal comes from the satellite towards Earth, it collides with Earth surface and some of them get absorbed. These are taken care by atmospheric absorption loss given by "AA" and measured in dB. Now, we can write the loss equation for free sky as,

$$\text{Losses} = \text{FSL} + \text{RFL} + \text{AML} + \text{AA} + \text{PL} \qquad \text{...(2.27)}$$

where,

$\quad$ RFL = stands for received feeder loss and unit is dB.

$\quad$ PL = stands for polarization mismatch loss in dB.

Now the decibel equation for received power can be written as,

$$P_R = EIRP + G_R + Losses \qquad \qquad ...(2.28)$$

where,

G_R = Receiver antenna gain.

This is a complete link design equation for satellite communication.

2.5 EARTH (GROUND) STATION

2.5.1 Introduction

- Earth station may be located on the surface of the Earth or in its atmosphere. They communicate with spacecraft by transmitting and receiving radio waves in the super high frequency (SHF) or extremely high frequency (EHF) bands e.g., microwaves. When a ground station successfully transmits radio waves to a *spacecraft* or vice versa, it establishes a telecommunications link. A principal telecommunications device of the ground station is the parabolic antenna.

- Earth stations may have either a fixed or itinerant position. Specialized satellite Earth stations are used to telecommunicate with satellites. Other ground stations communicate with manned *space stations* or unmanned *space probes*. A ground station that primarily receives telemetry data, or that follows a satellite not in *geostationary orbit*, is called a *tracking station*.

- When a satellite is within an Earth's *line-of-sight* (LoS), the station is said to have a view of the satellite. It is possible for a satellite to communicate with more than one ground station at a time. A pair of ground stations are said to have a satellite in *mutual view* when the stations share simultaneous, unobstructed, *line-of-sight contact* with the satellite.

2.5.2 Definition

- **A ground station (or Earth station) is a terrestrial radio station designed for extraplanetary telecommunication with spacecraft or reception of radio waves from astronomical radio sources.**

- *An Earth station is a collection of equipments installed on the Earth's surface that enables communications over one or more satellite.*

2.5.3 Block Diagram

- Designing of an Earth station depends not only on the location of Earth station but also on some other factors. The location of Earth stations could be on land, on ships in sea and on aircraft.

- The depending factors are type of service providing, frequency bands utilization, transmitter, receiver and antenna characteristics.

- The block diagram of digital Earth station is shown in Fig. 2.2.

2.5.4 Earth-Space Propagation Effect

- In space communication, with an Earth station as one terminal, several problems rise due to refraction, absorption and scattering effects, especially at microwave frequencies.

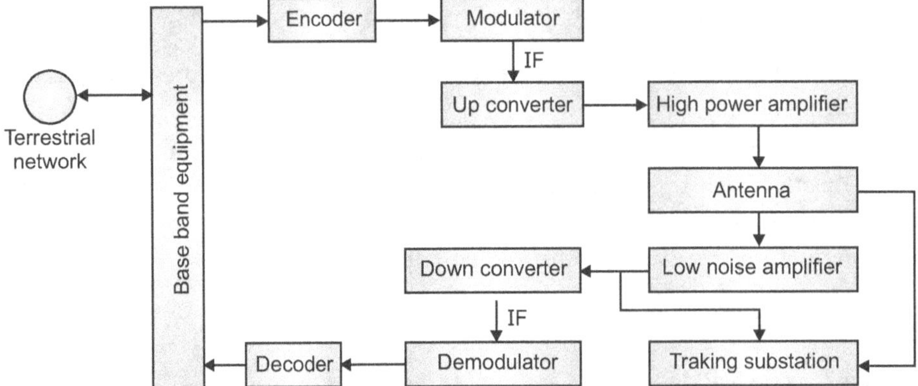

Fig. 2.2 : Digital Earth station

- For low angles of elevation of the Earth station beam, it is often necessary to evaluate the refraction produced by the troposphere, i.e. to determine the error in observed location of a satellite. The major part of the bending occurs in the first two kilometres above ground and some statistical correlation exists between the magnitude of the effect and the refractive index at the surface.

- This phase distortion also affects the stability of frequencies transmitted through the troposphere. Absorption in clear air may affect the choice of frequencies, above about 40 GHz, to minimize co-channel interference. Clouds produce an additional loss which depends on their liquid-water content. Layer-type cloud (stratocumulus) will not cause additional attenuation of more than about 2 dB, even at 140 GHz.

- Water drops attenuate microwaves both by scattering and by absorption. Scatter from rain (and ice crystals at and above the freezing level in the atmosphere) can cause significant interference on co-channel terrestrial and space systems even when the beams from the two systems are not directed towards each other on a great-circle path.

- Because precipitation (and to a smaller extent atmospheric gases) absorb microwaves, they also radiate thermal-type noise. Absorption in precipitation becomes severe at frequencies above about 30 GHz and **scintillation** effects also increase in importance in the millimeter range.

2.5.5 Interference Effects

- Interference effects are of utmost importance to the reliable design of modern satellite communication system operating at Ku and Ka bands. In these frequencies, rain attenuation is the dominant *fading mechanism* particularly for Earth-space systems located in subtropical and tropical regions. On the other hand, the main propagation effect on interference between two adjacent satellite systems is considered to be the differential rain attenuation.

- FDMA uses frequency allocation among a set of Earth station to share a series of frequencies. The transponder bandwidth is divided into a series of nonoverlapping frequency slots, each of which constitutes an access channel. Because the nonlinearity of the satellite transponder causes intermodulation products, nonlinear signal power transfer, and intelligible crosstalk, the TWTA cannot be operated to full capacity. Therefore, the power efficiency of an FDMA system is decreased because of the need to decrease the power of the TWTA.

- The limits on the power radiated from Earth station at low angle of elevation in frequency bands shared with the fixed service. These limits are a sharing constraint, designed to keep within reasonable bounds the distances at which interference to terrestrial stations can occur. It has much impact on the design of modern satellite networks. The interference distances should be relatively short. Such interference may be sporadic, depending, for example, on tropospheric duct propagation.

2.5.6 Earth Station Parameters

- Certain Earth station requirements are considered mandatory. Such parameters are categorised in this way in order to guarantee the protection and performance of the SES system and their customers, or to meet internationally agreed standards.

  1. Uplink and downlink frequencies
  2. Full band frequency ranger
  3. Frequency resolution
  4. Satellite transponder translation frequency tolerance
  5. Inclined orbit space work
  6. Transmit and receive cross polarization isolation
  7. Transmit co-polar and cross-polar side lobes
  8. Antenna receive co-polar and cross polar pattern
  9. Pointing stability
  10. Linear polarisation angle
  11. Transmission time delay variations
  12. Uplink EIRP adjustment
  13. Uplink power spectral density limits
  14. EIRP stability
  15. In-band emissions
  16. Out of band emissions
  17. Intermodulation products
  18. Beacon frequencies

2.6 FREQUENCY WINDOW

2.6.1 Concept

- When you use Fast Fourier Transform (FFT) to measure the frequency component of a signal, you are basing the analysis on finite set of data. The actual FFT transform assume that it is a finite data set, a continuous spectrum that is one period of a periodic signal.

- For the FFT, both the time domain and frequency domain are circulator topologies. So, the two endpoints of the time waveform are interpreted as though they were connected together. When the measured signal is periodic and an integer number of periods fill the acquisition time interval, then FFT turns out fine as it matches this assumption.

- However, many times, the measured signal is not an integer number of periods. Therefore, the finiteness of the measured signal may result in a truncated waveform with different characteristics from the original continuous-time signal, and the finiteness can introduce sharp transition changes into the measured signal. The sharp transitions are discontinuous.

- When the number of periods in the acquisition is not an integer, the endpoints are discontinuous. These artificial discontinuities show up in the FFT as high-frequency components not present in the original signal. These frequencies can be much higher than the *Nyquist frequency* and are aliased between 0 and half of your *sampling rate.*

- The spectrum you get by using a FFT, therefore, is not the actual spectrum of the original signal, but a smeared version. It appears as if energy at one frequency leaks into other frequencies. This phenomenon is known as *spectral leakage*, which causes the fine spectral lines to spread into wider signals.

- You can minimize the effects of performing an FFT over a noninteger number of cycles by using a technique called *windowing*. Windowing reduces the amplitude of the discontinuities at the boundaries of each finite sequence acquired by the digitizer.

- Windowing consists of multiplying the time record by a finite-length window with an amplitude that varies smoothly and gradually towards zero at the edges. This makes the endpoints of the waveform meet and, therefore, results in a continuous waveform without sharp transitions. This technique is also referred to as applying a window.

2.6.2 Functions

- There are several different types of *window functions* that you can apply depending on the signal. To understand how a given window affects the frequency spectrum, you need to understand more about the frequency characteristics of windows.

- An actual plot of a *window* shows that the frequency characteristic of a window is a continuous spectrum with a main lobe and several side lobes. The main lobe is centered at each frequency component of the time-domain signal, and the side lobes approach zero. The height of the side lobes indicates the affect the windowing function has on frequencies around main lobes.

- The *side lobe* response of a strong sinusoidal signal can overpower the main lobe response of a nearby weak sinusoidal signal. Typically, *lower side lobes* reduce leakage in the measured FFT but increase the bandwidth of the major lobe. The side lobe *roll-off rate* is the asymptotic decay rate of the side lobe peaks. By increasing the side lobe roll-off rate, you can reduce spectral leakage.

- Selecting a window function is not a simple task. Each window function has its own characteristics and suitability for different applications. To choose a window function, you must estimate the frequency content of the signal.

2.7 SATELLITE SUBSYSTEMS

2.7.1 Introduction

- In satellite communication system, various operations take place. Among which the main operations are orbit controlling, altitude of satellite, monitoring and controlling of other subsystems.

2.7.2 Types

- A satellite communication consists of mainly two *segments*. Those are *space segment* and *Earth segment*. So, accordingly there will be two types of subsystems as under:

  1. Space segment subsystems

  2. Earth segment subsystems.

1. Space Segment Subsystems:

- The subsystems present in space segment are called as space segment subsystems.

- Following are the space segment subsystems:

  1. Altitude and Orbital Control Subsystem

  2. Telemetry, Tracking, Command and Monitoring Subsystem

  3. Power and Antenna Subsystem

  4. Transponders

2. Earth Segment Subsystems:

- The subsystems present in the ground segment have the ability to access the satellite repeater in order to provide the communication between the users. Earth segment is also called as ground segment.

- Earth segment performs mainly two functions. Those are transmission of a signal to the satellite and reception of signal from the satellite. Earth stations are the major subsystems that are present in Earth segment.

2.7.3 Telemetry and Monitoring Subsystem

- The word *'Telemetry'* measurement at a distance mainly, the following operations take place in Telemetry:

  1. Generation of an electrical signal, which is proportional to the quantity to be measured.

  2. Encoding the electrical signal.

  3. Transmitting this code to a far distance.

- Telemetry subsystem present in the satellite performs mainly two functions :

  1. Receiving data from sensors,

  2. Transmitting that data to an Earth station.

- Satellites have quite few sensors to monitor different parameters such as pressure, temperature, status etc., of various subsystems. In general, the telemetry data is transmitted as FSK or PSK.

- Telemetry subsystem is a remote controlled system. It sends monitoring data from satellite to Earth station. Generally, the telemetry signals carry the information related altitude, environment and satellite.

2.7.4 Tracking Subsystem

- Tracking subsystem is useful to know the position of the satellite and its current orbit. Satellite Control Center (SCC) monitors the working and status of space segment subsystems with the help of telemetry downlink and, it controls those subsystems using command uplink.

- All satellites are tracked using two main indicators called *Azimuth* and *Elevation*. These two measurements have been used since the first day, when man started writing down the partitions of the *stars* and *planets* as observed from the Earth.

- We know that the *tracking subsystem* is also present in an Earth station. It mainly focusses on range and look angles of satellite. Number of techniques that are using in order to track the satellite. The change in the orbital position of satellite can be identified by using the data obtained from velocity and acceleration sensors that are present on satellite.

- The tracking subsystem that is present in an Earth station keeps tracking of satellite, when it is released from last stage of *Launch vehicle*. It performs functions like locating of satellite in *initial orbit* and *transfer orbit*.

2.7.5 Commanding Subsystem

- The satellite must form the satellite operations center (SOC) what its current stating and where it is located in orbit. Often a simple *Beacon* system is used to allow the Earth station track the satellite in orbit.

- Additional information is relayed to the ground such as the craft's operating temperature state of its programs and operating system as well as a host of other internal functions.

- Commanding subsystem is necessary in order to launch the satellite in an orbit and its working in that orbit. This subsystem adjusts the altitude and orbit of satellite, whenever there is a deviation in those values. It also controls the communication subsystem. This commanding subsystem is responsible for turning ON / OFF of other subsystems present in the satellite based on the data getting from telemetry and tracking subsystems.

- In general, *control codes* are converted into *command words*. These command words are used to send in the form of TDM frames. Initially, the validity of command words is checked in the satellite. After this, these command words can be sent back to Earth station. Here, these command words are checked once again.

- If the Earth station also receives the same (correct) command word, then it sends an execute instruction to satellite. So, it executes that command.

- Functionality wise, the Telemetry subsystem and commanding subsystem are opposite to each other. Since, the first one transmits the satellite's information to Earth station and second one receives command signals from Earth station.

Practice Questions

1. Derive satellite link design equation.
2. What is system noise temperature?
3. What are C/N ratio and G/T ratio?
4. Explain atmospheric and ionospheric effects on satellite link design.
5. Derive complete link design system for satellite communication.
6. Explain interference effects on complete link design.
7. List Earth station parameters.
8. Draw block diagram of Earth station.
9. What are Earth-space propagation effects?
10. Write short notes on the following :

    (a) Free space, (b) Atmospheric absorption, (c) Rain-fall attenuation, and (d) Ionospheric scintillations
11. Explain the concept of frequency window.
12. What are the functions of frequency window?
13. List subsystems of satellite communication.
14. Write short note on the following satellite subsystems:

    (a) Telemetry and monitoring, (b) Tracking, and (c) Commanding

SATELLITE MULTIPLE ACCESS SYSTEMS

FDMA techniques, SCPC and CSSB systems. TDMA frame structure, Burst structure, Frame efficiency, Super-frame, Frame acquisition and synchronization. TDMA versus FDMA, Burst time plan, Beam hopping. Satellite switched, Erlang call congestion formula, DA-FDMA, DA-TDMA.

3.1 FDMA TECHNIQUE

3.1.1 Introduction

- Frequency division multiple access technique (FDMA) is one of the most common analogue multiple access techniques.

- **Frequency Division Multiple Access (FDMA) is a channel access technique found in multiple-access protocols as a channelization protocol.**

3.1.2 Description

- The acronym FDMA stands for Frequency Division Multiple Access.

- The frequency band is divided into channels of equal bandwidth so that each conversation is carried on a different frequency.

- In FDMA method, *guard bands* are used between the adjacent signal spectra to minimize crosstalk between the channels. A specific frequency band is given to one person, and it will be received by identifying each of the frequency on the receiving end. It is often used in the first generation (1G) of analog mobile phone.

- FDMA is a channel access method used in some multiple-access protocols. It allows multiple users to send data through a single communication channel, such as a coaxial cable or microwave beam, by dividing the bandwidth of the channel into separate non-overlapping frequency sub-channels and allocating each sub-channel to a separate user. Users can send data through a subchannel by modulating it on a carrier wave at the subchannel's frequency.

- FDMA uses low bit rates (large symbol time) as compared to average delay spread.

3.1.3 Working Principle

- One way to understand FDMA is to imagine different people in the same class room communicating in voice with different *pitches*, some high and some low.

- They would be able to talk simultaneously and more or less understand one another. This is similar to the way FDMA works. It is used traditionally as AM and FM radio bands to allow broadcast by individual stations.

- FDMA is implemented at the media access control (MAC) layer of the data-link layer in the Open System Interconnection (OSI) reference model for networking protocol stations.

- FDMA is based on the Frequency-Division Multiplexing (FDM) technique used in wireless networking. In FDMA, the user is assigned a specific frequency band in the *electromagnetic spectrum*, and during a call that user is the only one who has the right to access the specific band.

- Two different frequency bands are used to allow full-duplex communication between base and mobile stations. Both of these bands are then divided into discrete channels that are 30 kHz wide in bandwidth.

3.1.4 Advantages

1. It reduces the bit rate information.
2. It increases the capacity due to use of efficient numerical codes.
3. It reduces the cost and lowers the inter symbol interference (ISI)
4. Equalization is not necessary.
5. It can be easily implemented.
6. It can be easily configured.
7. It improves speech encoder.
8. It easily reduces bit data.
9. It requires less number of bits for synchronization and framing.

3.1.5 Disadvantages

1. The maximum flow rate per channel is fixed and small.
2. It does not differ significantly from analog system.
3. Improvement in capacity depends on the signal-to-interference reduction, or a signal-to-noise ratio.
4. Guard band leads to wastage of capacity.
5. It cannot be realized in VLSI due to the use of narrowband filters.

3.1.6 Applications

1. Satellite communication
2. Telephone trunk lines

3.1.7 Types

1. Multi-channel per-carrier (MCPC)
2. Single-channel per-carrier (SCPC)

3.2 SINGLE CHANNEL PER CARRIER (SCPC) SYSTEM

3.2.1 Definition

- **SCPC is FDMA techniques having a single channel, i.e. either voice channel or data channel per RF carrier a satellite transporter.**

3.2.2 Description

- The abbreviation SCPC stands for single channel per carrier.
- Single channel per carrier (SCPC) is a widely used term in satellite communications. SCPC means single channel, i.e. either voice channel or data channel per RF carrier in a satellite transponder.
- As it is imperative from SCPC concept that only single channel is occupying the carrier all the time, hence it is not an efficient system in the use of satellite resource i.e. carrier for FDMA access assignment. Hence VOX feature is enabled to save power, which will only enable RF when voice activity is detected.
- Thus SCPC refers to using a single signal at a given frequency and bandwidth. Most often, this is used on broadcast satellites to indicate that radio stations are not multiplexed as subcarriers onto a single video carrier, but instead independently share a transponder.
- It may also be used on other communications satellites, or occasionally on non-satellite transmissions.

3.2.3 Advantages

1. Simple and reliable technique.
2. Low-cost equipment.
3. Any bandwidth upto a full transponder, usually 64 kbps to 50 Mbps.
4. Easy to add additional receiver sites i.e. Earth stations.

3.2.4 Disadvantages

1. Inefficient use of satellite bandwidth for burst transmissions, typically encountered with packet data transmission.
2. Usually it requires on-site control.
3. In remote locations, the transmitting dish antenna must be protected.

3.2.5 Applications

1. It is used on broadcast satellites to indicate radio stations are not multiplexed.
2. It is used on communication satellites, or on non-satellite transmissions.
3. It is used in satellite radio for continuous broadcast.
4. It is used in voice, where a small amount of fixed bandwidth is required.

3.3 COMPANDED SINGLE SIDE BAND (CSSB) SYSTEM

3.3.1 Definition

- The abbreviation CSSB stands for Companded Single Side Band.
- Companded single-side band (CSSB) is a narrowband modulation method using a single-sideband with a pilot tone, allowing an expander in the receiver to restore the amplitude that was severely compressed by the transmitter.
- **A long haul microwave telecommunication system that employ repeaters and single side band amplitude. Modulation and achiever subjective noise improvement by confiding to reduce circuit noise between syllables and during pauses in speech is called CSSB system.**

3.3.2 Description

- *CSSB system offers improved effective range over standard SSB modulation, while* simultaneously retaining backwards compatibility with standard SSB radios. It also offers reduced bandwidth and improved range for a given power level compared with narrow band FM modulation.
- The *companding* used in amplitude CSSB is a type of dynamic range reduction, wherein the difference in amplitude between the louder and softer sounds is reduced prior to transmission. A corresponding expander circuit in the receiver inverts this transformation in order to restore the dynamic range. If a conventional SSB receiver is used to receive ACSSB signals, some distortion may be noticed, but generally the signals are quite intelligible. Similar techniques are used in audio noise reduction circuits such as those developed for Dolby.
- A CSSB is being used by amateur radio operators, air-to-ground phones, as well as mobile-satellite services.

3.3.3 Applications

1. It is used in dolby system.
2. It is used in amateur radio receivers.
3. It is used to reduce audio noise in amplifiers.
4. It is used in air-to-ground phones.
5. It is used for mobile satellite services.

3.4 TDMA TECHNIQUES

3.4.1 Definition of TDMA

- The abbreviation TDMA stands for Time Division Multiple Access.
- **Time Division Multiple Access (TDMA) is a digital cellular telephone communication technology.**
- It facilitates many users to share the same frequency without interference. Its technology divides a signal into different timeslots, and increases the data carrying capacity.

3.4.2 TDMA Overview

- TDMA is a complex technology, because it requires an accurate synchronization between the transmitter and the receiver. It is used in digital mobile radio systems. The individual mobile stations cyclically assign a frequency for the exclusive use of a time interval.

- In most of the cases, the entire system bandwidth for an interval of time is not assigned to a station. However, the frequency of the system is divided into sub-bands, and TDMA is used for multiple access in each sub-band. Sub-bands are known as ***carrier frequencies***. The mobile system that uses this technique is referred to as the ***multi-carrier system***.

3.4.3 Advantages

1. It permits flexible data rates.
2. It can withstand gusty or variable bit rate traffic.
3. No guard band required for the wideband system.
4. No narrow band filter required for the wideband system.

3.4.4 Disadvantages

1. It requires complex quantization for high data rates of broadband systems.
2. It requires a large number of additional bits for synchronization and supervision due to the burst mode.
3. It needs call time in each slot to accommodate time to inaccuracies due to clock instability.
4. It increases energy consumption due to electronics operating at high bit rates.
5. It requires complex signal processing to synchronize within short slots.

3.5 TDMA FRAME STRUCTURE

3.5.1 Introduction

- Each user is allowed to transmit only within the specified time intervals (time-slots). Different users transmit in different time-slots. When users transmit, they occupy the whole frequency bandwidth, the separation among users is performed in the time domain.

3.5.2 Features

1. TDMA shares a single carrier frequency with several users, where each users makes use of non-overlapping time-slots.
2. The number of time-slots per frame depends on several factors, such as modulation technique, available bandwidth, etc.
3. Data transmission for users of a TDMA system is not continuous, but occurs in bursts.
4. The hands-off process is simpler for a subscriber unit, since it is able to listen for other base stations during idle time-slots.
5. TDMA uses different time-slots for transmission and reception. Thus, duplexers are not required.
6. Adaptive equalization is usually necessary in TDMA systems, since the transmission rates are generally very high as compared to FDMA channels.
7. In TDMA, the guard time should be minimized.
8. High synchronization overhead is required in TDMA systems, because of burst transmissions.
9. TDMA systems have overheads as compared to FDMA systems.

3.5.3 TDMA Frame Efficiency

- The frame efficiency is the percentage of bits per frame which contain transmitted data.
- Frame efficiency is a measure of the percentage of transmitted data that contains information as opposed to providing overhead for the access scheme.
- The transmitted data may include source and channel coding bits, so the efficiency of a system is generally less than frame efficiency.

3.5.4 Description

- TDMA requires a centralized control node, whose primary function is to transmit a periodic reference burst that defines a frame and forces a measure of synchronization of all the users. The frame so-defined is divided into time-slots, and each user is assigned a time-slot in which to transmit its information.

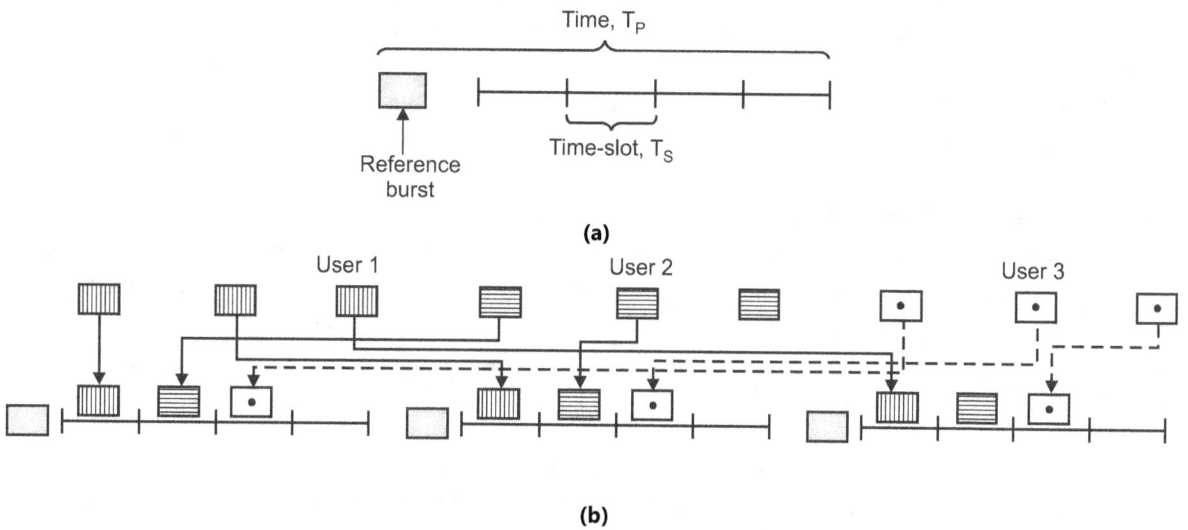

Fig. 3.1 : TDMA frame structure

3.6 BURST IN TDMA

3.6.1 Introduction

- **Burst is a term used in a number of information technology context to mean a specific amount of data sent or received in one intermittent operation.**

- It can be contrasted with streamed, paced, or continuous. Generally, a burst operation implies that some threshold has been reached that triggers the burst. Depending on particular technology, a burst operation can be intermittent at a regular or an irregular rate.

3.6.2 Definition

- **The data transmitted during single time-slot is known as a burst.**

- In telecommunication, a **burst transmission** or **data burst** is the broadcast of a relatively high-bandwidth transmission over a short period.

- Burst transmission can be intentional, broadcasting a compressed message at a very high data signaling rate within a very short transmission time.

3.6.3 TDMA Burst Structure

- Fig. 3.3 shows a burst, whose time length may be 180 mS. At the start of the burst, the bit sequence follows a predetermined ***burst preamble or header*** designed to help the hub receiver demodulator lock onto the carrier. This is described as the *"carrier and bit timing recovery sequence and unique word"*.

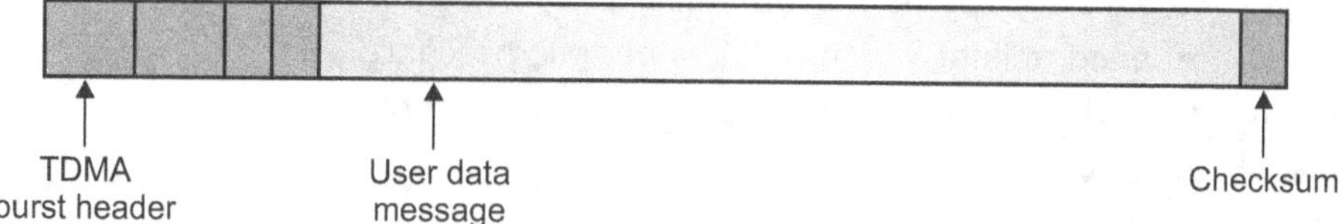

Fig. 3.2 : TDMA burst format

- A brief period of unmodulated CW carrier is a good start. This helps the demodulator Automatic Frequency Control (AFC) adjust its centre frequency and also get the Automatic Gain Control (AGC) so that the level of the carrier is made nominal.

- Next a period of alternate ones and zeros helps the demodulator to set the exact bit rate e.g. 249,999 bit/sec. Following this is the unique word. This is a pre-selected complex sequence of ones and zeros which the receiver attempts to match against two opposite matching patterns.
- When a hit is obtained the demodulator knows two things. It can now distinguish a one from a zero and it knows its exact position in the burst. At the end of the unique word the **user data message** starts. Right at the end may be a **cyclic redundancy (CRC) checksum** to check if there have been any error during the burst.
- Additional bits for **Forward Error Correction (FEC)** are also normally added. These may be distributed along the packet or added as a extra group of bits at the end. The CRC checksum simply detects if there have been any errors. FEC will put right and repair errors, provided there are not too many.

3.6.4 TDMA Burst Time Plan

- The allocation of burst time-slots within the TDMA frame is the burst time plan. Fig. 3.3 shows a sequence of two successive TDMA frames passing through the satellite. The carrier bit rate is 250 kbps.

From site 1	From site 2	From site 3	From site 4	From site 5	From site 1	From site 2	From site 3	From site 4	From site 5
0 180 mS	80 mS	180 mS	280 mS	180 mS	180 mS	80 mS	180 mS	280 mS	180 mS

|←———————————— One TDMA frame = one second = 1000 mS ————————————→|

Fig. 3.3 : Fixed time plan structure

- Site-1 transmits a burst, starting at the beginning of each TDMA frame. The burst lasts 180 mS, so at a rate of 250 kbps site-1 sends 45 kbits per burst, or 45 kbps.
- Site-2 transmits a burst, timed to arrive at the satellite just after the end of burst 1. The second burst lasts 80 mS, so at a rate of 250 kbps. Fig. 3.3 shows a fixed time plan, where each VSAT has been allocated a predetermined portion of the total time.
- This is designed in 20 mS guard period between each burst. This allows for slight mistiming in the transmission of bursts.
  VSAT site 1 start 0 mS, time allocated 180 mS
  VSAT site 2 start 200 mS, time allocated 80 mS
  VSAT site 3 start 300 mS, time allocated 180 mS
  VSAT site 4 start 500 mS, time allocated 280 mS
  VSAT site 5 start 800 mS, time allocated 180 mS
- This information is broadcast to all sites, which then follow the timing instructions. This burst time plan might be applied unchanged for several days or weeks or it might be changed every few seconds or minutes according to the traffic demand.

3.6.5 Burst Timing

- The time of arrival of each burst at the satellite is critical and to get this right each site is told when to transmit.
- This instruction is based on the burst time plan start time value plus also a time delay based on the range of VSAT site to the satellite. *The range is initially calculated based on the latitude and longitude of the Earth station and the orbital position of the satellite.* The amount of time to allow for, due to range, is calculated using the speed of light, which is 300 m per microsecond.
- When the site is activated, the burst should appear in the correct place.

3.7 TDMA SUPERFRAME

3.7.1 Definition

- In telecommunications, superframe is a T_1 framing standard.
- **The pattern sent is 12 bit long, so every group of 12 frames is called a superframe.**
- **In TDMA mechanism, the channel bounded by a structure that consists of a number of time slots allocated by a base of a coordinator is called a superframe structure.**

3.7.2 Description

- In telecommunications, **superframe (SF)** is a T_1 framing standard. In 1970s it replaced the original T_1/D_1 framing scheme of the 1960s in which the framing bit simply alternated between zero and one.
- Superframe is sometimes called **D_4 Framing** to avoid confusion with single-frequency signaling. It was first supported by the D_2 channel bank, but it was first widely deployed with the D_4 channel bank.
- In order to determine where each channel is located in the stream of data being received, each set of 24 channels is aligned in a frame. The frame is 192 bits (8×24) long, and is terminated with a 193rd bit, the framing bit, which is used to find the end of the frame.
- In order for the framing bit to be located by receiving equipment, a predictable pattern is sent on this bit. Equipment will search for a bit which has the correct pattern, and will align its framing based on that bit. The pattern sent is 12 bits long, so every group of 12 frames is called a superframe. The pattern used in the 193rd bit is 100011 011100.
- Each channel sends two bits of call supervision data during each superframe using robbed-bit signaling during frames 6 and 12 of the superframe.
- More specifically, after the 6th and 12th bit in the *superframe pattern*, the least significant data bit of each channel (bit 8; T_1 data is sent big-endian and uses 1-origin numbering) is replaced by a "*channel-associated signalling*" bit (bits A and B, respectively).
- Superframe remained in service in many places through the turn of the century, replaced by the improved Extended Superframe (ESF) of the 1980s in applications where its additional features were desired.

3.7.3 Structure

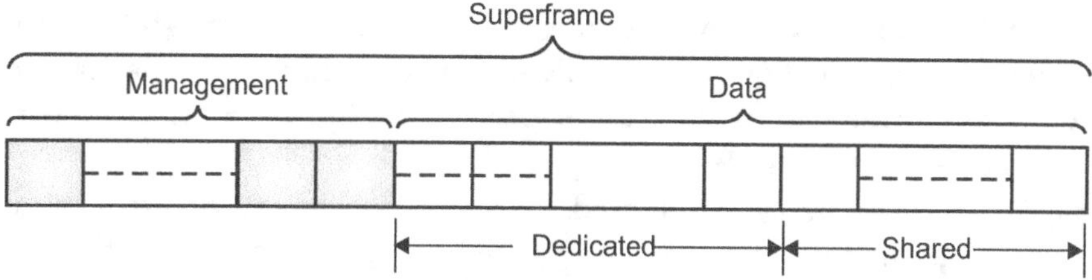

Fig. 3.4 : Superframe structure of TDMA

- The superframe is bounded by network beacons and divided into 16 equal sized time-slots. The first time-slot of each superframe is used to transmit the beacon. The main purpose of the beacon is to synchronize devices in the network, to identify the PAN, and to describe the superframe structure.
- The devices use a slotted CSMA/CA based protocol to gain access and compete for time-slots. All communications between devices must complete by the end of the current CAP and the beginning of the next network Beacon.

3.8 FRAME ACQUISITION AND SYNCHRONIZATION

3.8.1 Introduction

- Rapid advancements in wireless technology and evolution of a broad spectrum of applications have created a need to develop techniques and algorithms for precise operation of a satellite communication link. This necessitates an investigation of various components involved in a satellite communication. For example, channel sharing, data structure, signal specification, transmitter and receiver processing etc. are some of the areas.
- Multiple Earth stations can transmit data intermittently on the some frequency, due to which special have to be taken in order to avoid interference among data transmitted by different Earth stations. This is in addition to the Quality of Service (QoS) demands from the network.
- These requirements have created a critical importance for the development of efficient *acquisition* and *synchronization* algorithms. Moreover, the pace at which new applications based on satellite communications are being developed has resulted in rapid changes in satellite technology as well, which in turn has imposed additional constraints on *acquisition* and *synchronization* algorithms.

3.8.2 Frame Acquisition in TDMA

- With TDMA, only one carrier uses the transponder at one time and therefore the intermodulation products, which results from non-linear operation of amplifier working in multi carrier mode are absent in TDMA is suited for digital data transmission.

- Digital can be assembled into burst for transmission and reassembled from second bursts through use of digital buffer memories.

- Burst synchronization is required, and only one station is assigned solely for the purpose of transmission of bursts to which other stations are synchronized.

- *The time interval from the start of one reference burst to the next is termed as frame.* Frame contains the reference burst R and the bursts from the other Earth stations.

- Overall the transmission appears to be continuous because the input and output bit rates are continuous and equal. However, within the transmission channel, input bits are temporarily stored and transmitted in bursts.

- The basic units used in Earth station of TDMA system consist of multiplexer and demultiplexer units, preamble generator, phase modulator, etc. The reference bursts are required at the beginning of each frame to provide timing information for acquisition and synchronization of bursts.

- Acquisition refers to the process of positioning a *burst* into its assigned location in a *frame*.

3.8.3 Frame Synchronization in TDMA

- TDMA frame synchronization is important because of satellite motion and different propagation ranges which affect the time at which Earth stations must transmit so that their bursts do not overlap at the satellite.

- **In telecommunication, frame synchronization is the process by which, while receiving a stream of framed data, incoming frame alignment signals (i.e., a distinctive bitsequences or syncwords) are identified (i.e., distinguished from data bits), permitting the data bits within the frame to be extracted for decoding or retransmission.**

- If the transmission is temporarily interrupted, or a bit slip event occurs, the receiver must be synchronized. The transmitter and the receiver must agree ahead of time on which frame synchronization scheme they will use.

- In telemetry applications, a *frame synchronizer* is used to frame-align a serial pulse code-modulated (PCM) binary stream. The frame synchronizer immediately follows the bit synchronizer in most telemetry applications. Without frame synchronization, demodulation is impossible. The frame synchronization pattern is known as binary pattern which repeats at a regular interval within the PCM stream. The frame synchronizer recognizes this pattern and aligns the data into minor frames or sub-frames.

- Typically the frame sync pattern is followed by a counter (sub-frame ID) which dictates which minor or sub-frame in the series is being transmitted. This becomes increasingly important in the demodulation stage where all data is deciphered as to what attribute was sampled. Different commutations require a constant awareness of which section of the major frame is being decoded.

3.9 BEAM HOPPING TECHNIQUE

3.9.1 Concept

- Multi-beam satellite systems at Ka Band can provide broadband high capacity access to interactive multimedia services with a good compromise between complexity and cost. It is also well known that the traffic profile tends to be asymmetric from the user point of view, with dynamic behaviour increasing daily and higher levels of QoS being demanded.

- For this reason, means to increase system efficiency in support of the demanding dynamic traffic conditions are a key aspect to increasing system competitiveness whilst keeping its quality. On the other hand, the DVB-S2 standard is recognized as providing adaptability to propagation conditions, supporting higher transmission efficiency and dedicating more resources (bandwidth and/or power) to the users able to use them.

- In order to cope with this situation, the present project is intended to investigate a solution based on *beam hopping techniques*, consisting of the illumination of only a subset of the satellite beams through an appropriately designed beam illumination pattern.

- The challenge is to satisfy as close as possible the beam traffic requests, maximizing revenues through provision of high capacity in those areas of the coverage region with high demand, and minimizing the amount of resources dedicated to low loaded regions. As such, the focus on this activity is to fulfil the following objectives.

3.9.2 Objectives

1. Deeply investigate and assess the potential of beam hopping at system and payload levels.

2. Define payload architectures suitable for beam hopping systems.

3. Study, analyse and optimize beam hopping techniques in multibeam broadband satellite systems. The benefits in terms of capacity and system flexibility compared to conventional system architectures shall be assessed through detailed system simulations.

4. Compare the achieved performance with the one provided under the same conditions by system configurations, where no beam hopping, but an ad-hoc bandwidth/power allocation is utilized.

3.9.3 Features

1. To evaluate the improvements of system level provided by the use of beam hopping techniques, it is intended to analyse several schemes (referenced as study cases) and compare them in terms of capacity, performance and flexibility.

2. An appropriate optimization methodology for the complex problem to be solved will be defined. Then such an optimization process will be simulated and results will be iteratively obtained.

3.9.4 Optimization Methodology Steps

- The optimization methodology consists of the following steps:

1. A precise detailed system model will be obtained and the key parameters and interdependencies will be identified. In addition, the constraints given by the payload design will be considered in the optimization.

2. A pre-analysis with simplified models will be carried out, performing optimization upon identified key sub-problems, for example by setting other parameters as fixed.

3. A global optimization shall be applied, whose output shall be delivered to the system simulator in order to start an iteration that will yield the pursued optimal solution.

3.9.5 Optimized Solution

- Finally, the system simulation will use the optimization methodology to end up with an optimised solution.

1. An optimization algorithm module that transmits a set of optimized parameters to the system simulator, which has to return a value of the performance metric corresponding to the simulated satellite system.

2. The value of this performance metric will be used by the optimization algorithm to derive another set of optimized parameters.

3. A system simulator module, that will realistically model the long term performance of a satellite system for the two envisaged scenarios and for all study cases.

4. The results from the optimization process will be the selection of the optimum values for the identified parameters that determine the design of the beam hopping system. The possible impact of such a system on current standards, both at physical layer and MAC layer will be evaluated.

3.10 SATELLITE SWITCHED TDMA

3.10.1 Introduction

- The basic concept of satellite switched TDMA (SS/TDMS) was proposed already more than 20 years ago. Recognizing the effectiveness of this technology, *Takuro Muratani, Yasuhiko Ito* and *Takeshi Mizuike* played a pioneering role in development of this technology to finally implement a commercial system through their continuing effort.

- *Takuro Muratani* and *Yasuhiko Ito* first invented a solution algorithm to derive the most efficient sequence of switching pattern for given traffic demands among multiple spot beams.

- *Takuro Muratani, Yasuhiko Ito* and *Takeshi Mizuike* continued a pioneering work to prove and demonstrate feasibility of essential technical subjects through development of a prototype on-board satellite switch matrix to implement SS/TDMA system, redundancy configuration and diagnostic techniques and synchronization control mechanism between TDMA reference stations and satellite switching. They played a leading role in development of the specification of INTELSAT SS/TDMA system, for which they also conducted a research on an operation planning method and developed an original control algorithm to take maximum advantage of the SS/TDMA system.

- As a result of their effort and achievement, INTELSAT-VI satellite, which was equipped with on-board SS/TDMA dynamic switch matrix was developed in 1989.

3.10.2 Definition

- **Satellite Switched TDMA (SS/TDMA) is a highly efficient digital satellite communication system for multi-beam satellites.**

- SS/TDMA system realizes efficient satellite operation by dynamic switching of multiple satellite spot beams, which is synchronized with burst transmission signals to multiple destinations by TDMA.

3.10.3 Concept Description

- Fig. 3.5 shows a simplified form of the SS/TDMA concept developed by Scarcella and Abbott in 1983.

- Three antenna beams are used, each beam serving two Earth stations. A 3×3 satellite switch matrix is shown. This is the key component that permits the antenna interconnections to be made on a switched basis.

- A *switch mode* is a connectivity arrangement. With three beams, six modes would be required for full interconnectivity, as shown in Fig. 3.5 (*b*), and in general with N beams, $N!$ modes are required for full interconnectivity.

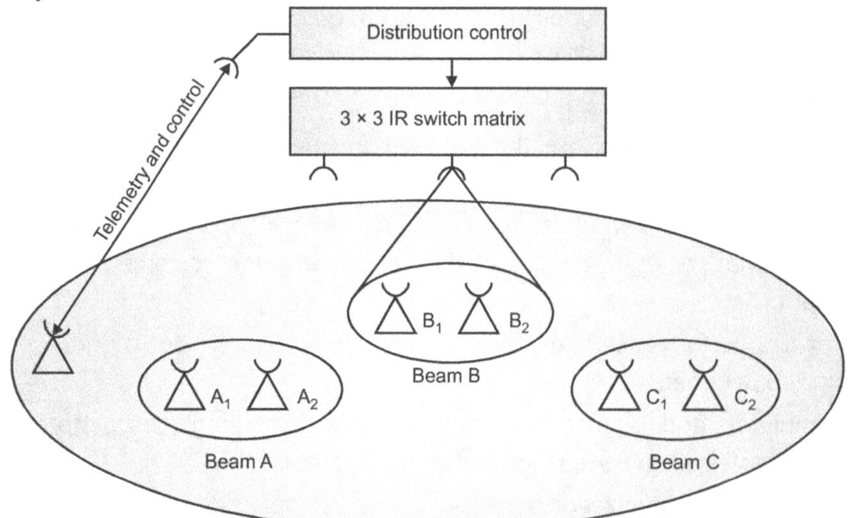

(a) Satellite switching of 3 spot beams

Input	Output					
	Mode 1	**Mode 2**	**Mode 3**	**Mode 4**	**Mode 5**	**Mode 6**
A	A	A	B	C	B	C
B	B	C	A	A	C	B
C	C	B	C	B	A	A

(b) Connectivities or modes

Fig. 3.5

- Full interconnectivity means that the signals carried in each beam are transferred to each of the other beams at some time in the switching sequence. This includes the loopback connection, where signals are returned along the same beam, enabling intercommunications between stations within a beam. Of course, the uplink and downlink microwave frequencies are different.

- Because of beam isolation, one frequency can be used for all uplinks, and a different frequency for all downlinks (e.g., 14 and 12 GHz in the Ku band).

3.11 COMPARISON OF TDMA AND FDMA

Sr. No.	Parameters	FDMA	TDMA
1.	Data rates	Low	Medium
2.	Mode of data	Continuous signal	Signal in bursts
3.	Capacity of the system	Low	Medium
4.	Cost	High	Low
5.	Handoff	Hard	Hard
6.	Flexibility	Low	Moderate
7.	Technique	Sharing of overall bandwidth of satellite transponder.	Sharing of time of the satellite transponder
8.	Synchronization	Not required	Essential
9.	Code word	Not required	Not required.
10.	Power efficiency	Reduced	Full power efficiency power
11.	Guard times and bands	Required	Required
12.	Interference effects	Adjacent frequency band interference.	Interference between user of adjacent time slots.
13.	Variable transmission rate	Difficult	Easy
14.	Near-far problem	No	No
15.	Fading mitigation	Equalizer not needed.	Equalizer may need.

3.12 ERLANG CALL CONGESTION FORMULA

3.12.1 Introduction

- The Erlang is named after a Danish telephone engineer named *A.K Erlang* (Agner Krarup Erlang). He was the first person to investigate traffic and queuing theory in telephone circuits.

3.12.2 What is an Erlang

- Erlang is a statistical measure of the voice traffic density in a telecommunication system.

- **An Erlang is the unit of measure of the voice traffic density in a telecommunication system or network and it is widely used for measuring load and efficiency.**

- It is necessary to understand the required capacity in a network to be able to provide correctly for it. Insufficient capacity and some calls cannot be carried, too much and this leads to excess costs for unused capacity.

- As a result it helps to have a definition of the telecommunications traffic so that the volume can be quantified in a standard way and calculations can be made.
- Telecommunications network designers make great use of the Erlang to understand traffic patterns within a voice network and they use the figures to determine the capacity that is required in any area of the network.

3.12.3 Erlang Definition

- **Erlang is a dimensionless unit that is used in telephony as a measure of offered load or carried load on service-providing elements such as telephone circuits or telephone switching equipment.**
- A single cord circuit has the capacity to be used for 60 minutes in one hour. Full utilization of that capacity, 60 minutes of traffic, constitutes 1 erlang.

3.12.4 Basics of Erlang

- The Erlang unit is the basic measure of telecommunications traffic intensity representing continuous use of one circuit and it is given the symbol "E".
- Erlang is effectively call intensity in call minutes per sixty minutes. In general the period of an hour is used, but it actually a dimensionless unit because the dimensions cancel out (i.e. minutes per minute).
- The number of Erlangs is easy to deduce in a simple case. If a resource carries one Erlang, then this is equivalent to one continuous call over the period of an hour.
- Alternatively if a radio channel is used for fifty percent of the time carries a traffic level of half an Erlang (0.5E). From this it can be seen that an Erlang, E, may be thought of as a use multiplier where 100% use is 1E, 200% is 2E, 50% use is 0.5E and so forth.

3.12.5 Erlang Formula

- **Erlang is a traffic modeling formula used in call center scheduling to calculate delays or predict waiting times for callers.**
- Erlang C bases its formula on three factors:
- The number of staff providing service, the number of caller waiting, and the average amount of time it takes to serve each caller.
- Erlang C can also calculate the resources that will be needed to keep wait times within the call center's target limits. This method assumes that there are no lost calls or busy signals, and therefore may overestimate the staff that is required.
- It is possible to express the way in which the number of Erlangs are required in the format of a simple function or formula.

$$E = \lambda \times h \qquad \qquad ...(3.1)$$

where:

λ = the mean arrival rate of new calls

h = the mean call length or holding time

E = the traffic in Erlangs.

- Using this simple Erlang function or Erlang formula, the traffic can easily be calculated.
- The use of the Erlang and the basic concepts surrounding its use have provided telecommunications engineers a valuable tool. It is widely used within the industry to look at loading levels, especially in areas like call centres, telephone exchanges and lines linking different areas.
- However in this basic form the Erlang does not address some real life aspects of loading including peak traffic density and the number of blocked calls resulting from short term overloading. To address these factors, developments of the basic Erlang concept have been introduced and they include measures like the Erlang B and Erlang C.
- The Erlang B formula, also known as the Erlang loss formula, is a formula for the blocking probability that describes the probability call losses for a group of identical parallel resources (telephone lines,

circuits, traffic channels, or equivalent). The formula is not limited to telephone networks but is also used in certain inventory systems with lost sales.

- The blocking probability of a bufferless loss system is given by,

$$P_b = B(E, m) = \frac{E^m/m!}{\sum\limits_{i=0}^{m} E^i/I!} \qquad \qquad ...(3.2)$$

where: m = Number of identical parallel resources

E = λh is the normalised ingress load

- The ***Erlang C formula*** expresses the probability that an arriving customer will need to queue as opposed to immediately being served. The Erlang C formula assumes that callers never hang up while in queue, the wait probability is given by,

$$P_w = \frac{\dfrac{E^m}{m!}\dfrac{m}{m-E}}{\left(\sum\limits_{i=0}^{m-1}\dfrac{E^i}{I!}\right) + \dfrac{E^m}{m!}\dfrac{M}{m-E}} \qquad \qquad ...(3.3)$$

where: E = total traffic

m = number of servers

3.12.6 Limitations

1. In the event of extremely high traffic congestion, Erlang's equation fail to accurately predict the correct number of circuits required because of re-entrant traffic. This is termed as high loss system.

2. If the service provider had not catered for this sudden peak demand, extreme traffic congestion will develop and Erlang's equations cannot be used.

3.13 DA-TDMA

3.13.1 Concept Description

- The abbreviation DA-TDMA stands for Demand Assigned-Time Division Multiple Access. DA-TDMA is a type of satellite access technology that is superior compared to FDMA or SCPC in terms of efficiency in satellite bandwidth usage.

- In DA-TDMA, satellite bandwidth is shared among the users at different sites based on allocation of time-slots rather than frequency. Hence, all the Earth stations of a DA-TDMA-based network will transmit at the same frequency, but not at the same time.

- In a DA-TDMA-based network, each Earth station is usually allocated a fixed percentage of satellite bandwidth (time-slots). Besides, there is a certain percentage of satellite bandwidth, which can be dynamically allocated to the Earth stations, based on users' demands.

- As a result, higher efficiency in the use of satellite bandwidth can be achieved. Due to their capability to dynamically allocate satellite bandwidth based on demand, DA-TDMA-based satellite networks are more suitable to carry *"bursty"* traffic than FDMA or SCPC-based satellite networks.

- The efficiency of such satellite bandwidth usage is achieved at the expense of timing delay. This is particularly so, as the *traffic* is using dynamically assigned satellite bandwidth that imposes a certain amount of delay.

- Such delay is proportional to the propagation delay of a typical satellite link.

3.13.2 Star Shaped DA-TDMA

- The Star Shaped DA-TDMA satellite network is as shown in Fig. 3.6. We can allocate K time-slots within a TDMA frame duration, to be shared by N VSATs at the satellite inbound transponder band.

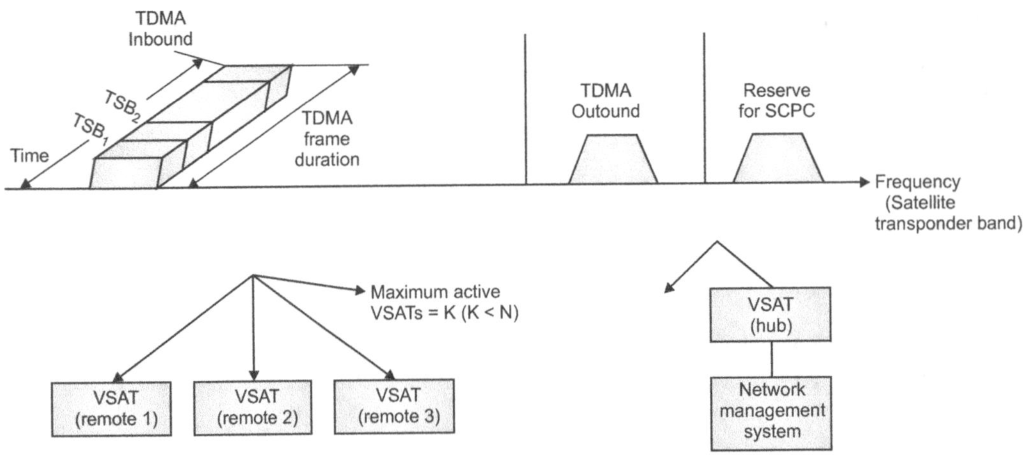

Fig. 3.6 : A Star Shaped DA-TDMA Inbound / TDM Outbound Satellite Network

- Every remote VSAT transmits its *carrier burst* at the same bandwidth and same frequency, but is not preassigned to transmit its carrier burst within a specific time-slot (that is the position and duration are not fixed). Any passive remote VSAT can request to setup a satellite link (by turning from carrier 'off' state to carrier 'on' state) with the hub VSAT, and access any unoccupied time-slot on the TDMA frame duration.

- Any active remote VSAT can also request the *Hub's Network* Management System to increase its capacity by extending the TDMA duration of its burst, to support a larger number of connections.

- When all the time-slots on the TDMA frame duration are being filled with *carrier bursts* (due to more traffic demand from remote VSATs), blocking of satellite link set-up may occur.

3.13.3 Implementation of DA-TDMA over Satellite Link

- There are three resources such as power, bandwidth and time, which are available to a satellite communication engineer. In demand-assigned TDMA technology, the satellite engineer achieves gains in bandwidth through more efficient use of satellite bandwidth, but at the expense of timing delay.

- This delay have an impact of the QoS on the various applications in which the demand-assigned TDMA-based satellite network is supporting. There are several timing-related issues involved in providing QoS over demand-assigned TDMA-based satellite networks as under :

  1. An increase in burstiness of the traffic of any particular source.

  2. The impact of the significantly increased delay on the performance of transport protocols, which require feedback mechanisms, for flow control.

  3. The actual implementation of QoS, which uses commercial-off-the-shelf (COTS) products, over demand-assigned TDMA-based satellite networks.

  4. The main considerations in implementing a DA-TDMA Satellite Network are as under:

     (i) Satellite bandwidth required to support a wide range of services.

     (ii) Choice of satellite modem.

     (iii) A sufficient energy-per-bit-to-noise density ratio i.e. Eb/No.

     (iv) Internet Protocol used, i.e. IPv4 and IPv6.

3.14 DA-FDMA

- The abbreviation DA-FDMA stands for Demand Assigned Frequency Division Multiple Access.

- The mobile satellite industry has reached a critical juncture. The mobile satellite industry has reached a critical juncture. Different potential service providers are buying for market shares and are planning their systems to meet market requirements with the most efficient designs. The procurement of satellites for the U.S. and Canadian MSSs is also imminent. With these developments, the debate over system architecture and multiple-access techniques has come into focus.

- The architecture proposed by JPL is based on FDMA. This choice of multiple-access scheme evolved out of system studies performed as early as 1983. In addition, the long propagation delay between the Earth and the geosynchronous satellite precludes the use of carrier sense type multiple access.

- To efficiently and effectively utilize the scarce resources of bandwidth and power, the approach of Demand Assigned Multiple Access (DAMA) is adopted for network implementation. Hence, the system architecture proposed by NASA/JPL utilizes a DA/FDMA scheme.

- In the demand-assigned mode of operation, the transponder frequency bandwidth is subdivided into a number of channels. A channel is assigned to each carrier in use, giving rise to the single-channel-per-carrier mode of operation discussed in the preceding section. As in the preassigned access mode, carriers may be frequency modulated with analog information signals, these being designated FM/SCPC, or they may be phase modulated with digital information signals, these being designated as PSK/SCPC.

- Demand assignment may be carried out in a number of ways. In the polling method, a master Earth station continuously polls all the Earth stations in sequence, and if a *call request* is encountered, frequency slots are assigned from the pool of available frequencies. The polling delay with such a system tends to become excessive as the number of participating Earth stations increases.

- Instead of using a polling sequence, Earth stations may request calls through the master Earth station as the need arises. This is referred to as *centrally controlled random access*. The requests go over a digital orderwire, which is a narrowband digital radio link or a circuit through a satellite transponder reserved for this purpose.

- Frequencies are assigned, if available, by the master station, and when the call is completed, the frequencies are returned to the pool. If no frequencies are available, the blocked call requests may be placed in a queue, or a second call attempt may be initiated by the requesting station.

Practice Questions

1. What is FDMA ?

2. Describe working principle of FDMA.

3. State advantages and disadvantages of FDMA.

4. List applications of FDMA.

5. What is SCPC ? Describe SCPC.

6. State advantages and disadvantages of SCPC.

7. List applications of SCPC.

8. What is CSSB ?

9. Describe CSSB.

10. List applications of CSSB.

11. What is TDMA ?

12. Describe TDMA.

13. What is time frame structure ?

14. State key features of TDMA frame structure.

15. What is TDMA and frame efficiency ?

16. What is burst in TDMA ?

17. Explain TDMA burst time plane.

18. What is burst timing ?

19. What is TDMA superframe ?

20. Describe TDMA superframe with its structure.

21. What is acquisition and synchronization ?

22. Explain frame acquisition in TDMA.

23. Explain frame synchronization in TDMA.

24. What is beacon hopping ?

25. Explain concept of beacon hopping.

26. What is satellite switched TDMA ?

27. Describe satellite switched TDMA.

28. Compare TDMA and FDMA.

29. What is Erlang and derive Erlang formula.

30. What are limitations of Erlang formula ?

31. What are DA-TDMA and DA-FDMA ?

32. Describe DA-TDMA.

☆☆☆

Chapter 4

SATELLITE SERVICES

INTELSAT, INSAT Series, VSAT, Weather forecasting, Remote sensing. LANDSAT, Satellite Navigation, Mobile Satellite Service.

4.1 INTELSAT

4.1.1 Origin

- *John F. Kennedy* the president of U.S. instigated the creation of INTELSAT with his speech to the United Nations on the 25th of September 1961. Less than a year later, *John F. Kennedy* signed the Communications Satellite Act of 1962. INTELSAT was originally formed as International Telecommunications Satellite Organization (ITSO) and operated from 1964 to 2001 as an *intergovernmental consortium* owning and managing a constellation of communications satellites providing *international broadcast services*. In 2001, the international satellite market was fully commercialized, and the US predominant role in INTELSAT was fully privatized after 2001 as INTELSAT was formed up as a private Luxembourg corporation.

- The International Government Organization (IGO) began on August 20, 1964; 55 years ago, with 7 participating countries. The 1964 agreement was an *interim arrangement* on a path to a more permanent agreement. The permanent international organization was established in 1973, following inter-nation negotiations from 1969 to 1971. The most difficult issue to *"resolve concerned the shift from management of the system by a national entity to management by the international organization itself."*

4.1.2 Brief History

- In 1945, *Arthur C. Clarke* published a paper in a magazine detailing his idea of a *'geostationary satellite'*. The geostationary orbit is known as the *Clarke orbit*. INTELSAT, the International Telecommunications Satellite Organization, which operates the global system, has started calling it *'Clarke orbit'*.

- A satellite in Clarke's orbit covers 44% of Earth's surface i.e. a large area. Moreover, due to its fixed relative position with a location on Earth, ground antennas would be simpler to build, thereby increasing the ease of communication.

- The famous *Sputnik* was the first artificial satellite to be launched into space. The satellite was launched into a low-Earth orbit by USSR on October 4, 1957. A low-Earth orbit has an altitude of 160-2000 km. Sputnik carried four external radio transmitters, and their pulses were detectable. The launch of Sputnik was a very significant milestone.

- The launch of Sputnik accelerated the development of technology and propelled U.S. to the top of the world concerning technical expertise. By the end of 1959, the U.S. had a satellite program to on the USSR.

- The SCORE was the first communication satellite, which was launched on December 18, 1958, aboard the U.S. Atlas rocket. With SCORE, a communication link was established for the first time in space.

- NASA launched the *'Telstar'* in 1962. It was a major technological breakthrough in terms of the types of data that could be communicated using a satellite. The Telstar was the first satellite to relay three different types of information. Telephone, Television, and high-speed data communications. Moreover, it provided the very first live transatlantic television feed. It is still in orbit.

- The *'Syncom 1'* was the first satellite to try to achieve a *geosynchronous orbit*. *'Syncom 2'* launched a few months later in 1963, was the first satellite to achieve a geosynchronous orbit. *'Syncom 3'* finally reached the geostationary orbit in 1964.

- On April 6, 1965, INTELSAT's first satellite, the INTELSAT 1 nicknamed *'Early Bird'* was geostationary orbit satellite above the all *Atlantic ocean* by a Delta D rocket. This satellite carry 240 telephone lines.

- The *'Nimbus 3'*, launched by NASA in 1969 was the first satellite for search and rescue operations.

- *'Anik 1'*, launched by Canada in 1972, became the first domestic communications satellite system using geosynchronous orbit.

- In 1978, the Indian Space Research Organization (ISRO) launched *'Aryabhatta'*, the first Indian satellite. Presently, ISRO has accomplished quite a few record breaking missions. Key among them was a mission to *Mars* and *Moon.*

- Only July 18, 2001, INTELSAT became private company 37 years after its formation. Today, the number of INTELSAT satellites, as well as ocean-spanning fiber-optic lines, allows rapid rerouting of traffic when one satellite fails. Modern satellites are more robust, lasting longer with a much larger capacity.

- As of 2018, INTELSAT provides services to over 600 Earth stations in more than 150 countries, territories and dependencies.

4.1.3 Definition

- INTELSAT stands for International Telecommunication Satellite Organization. It is a leading provider of communication satellite services which operates the global system.

4.2 INSAT SERIES

4.2.1 Definition

- INSAT stands for Indian National Satellite system. **INSAT is a series of multipurpose geostationary satellite launched by Indian Space Research Organization (ISRO) to satisfy the telecommunications, broadcasting, meteorology, and search and rescue operations.**

4.2.2 INSAT System

- INSAT is the largest domestic communication system in the Indo-Pacific region. It is a joint venture of an the Department of Space, Department of Telecommunications, Indian Meteorological Department, All India Radio and Doordarshan. The overall coordination and management of INSAT system rests with the Secretary-level INSAT Coordination Committee.

- INSAT system was commissioned with the launch of INSAT-1B in August 1983 (INSAT-1A, the first satellite was launched in April 1982, but could not fulfil the mission). INSAT system ushered in a revolution in India's television and radio broadcasting, telecommunications and meteorological sectors. It enabled the rapid expansion of TV and modern telecommunication facilities to even the remote areas and off-shore islands.

- INSAT satellites provide *transponders* in various bands (C, S, Extended C and Ku) to serve the television and communication needs of India. Some of the satellites also have the Very High Resolution Radiometer (VHRR) and Charge-Coupled Device (CCD) cameras for meteorological imaging.

- The satellites also incorporate transponder(s) for receiving distress alert signals for search and rescue missions in the *South Asian* and *Indian Ocean Region*, as ISRO is a member of the *Cospas-Sarsat program*. Satellites are monitored and controlled by Master Control Facilities (M.C.F.) that exist in *Hassan* and *Bhopal.*

4.2.3 Satellites in Service

- Of the 24 satellites launched in the course of the INSAT program, 11 are still in operation.

   1. **INSAT-2E:** It is the last of the six five satellites in INSAT-2 series Prateek. It carries seventeen C-band and lower extended C-band transponders providing zonal and global coverage.

   2. **INSAT-3A:** The multipurpose satellite, INSAT-3A, was launched by *Ariane* in April 2003.

   3. **INSAT-3C:** It was launched in January, 2002. All the transponders provide coverage over India.

   4. **INSAT-3D:** It was launched in July, 2013. All the transponders provide coverage over India, Bangladesh, Bhutan, Maldives, Nepal, Seychelles, Sri Lanka and Tanzania.

   5. **INSAT-3DR:** It is a *weather satellite* and was launched on September 9, 2016. It is a follow-up to INSAT-3D.

   6. **INSAT-3E:** It was launched in September, 2003. It has been decommissioned and gone out of service from April 2014. GSAT-16 will replace this satellite.

   7. **KALPANA-1:** It is an exclusive *meteorological satellite* launched by PSLV in September 2002. Its first name was METSAT. It was later renamed as KALPANA-1 to commemorate Kalpana Chawla.

   8. **INSAT-4 Series:**

      (a) **INSAT-4A:** It was launched in December 2005 by the European Ariane launch vehicle. It was used by TATA Group and STAR uses INSAT-4A for distributing their DTH service.

      (b) **INSAT-4B:** It was launched in March 2007 by the European Ariane launch vehicle. It has 12 transponders in the C band used for television, radio and telecommunication purposes.

      (c) **INSAT-4CR:** It was launched on 2 September 2007 by GSLV-F04. It is a replacement satellite of INSAT-4C which was lost when GSLV-F02 failed and had to be destroyed. It incorporates a Ku band Beacon as an aid to tracking the satellite.

   9. **GSAT Series:** The GSAT satellites are India's indigenously developed communications satellites, used for *digital audio, data* and *video broadcasting* for both military and civilian users. As of November 2018, 19 GSAT satellites of ISRO have been launched out of which 15 satellites are currently in service.

      They are GSAT-2, GSAT-3, GSAT-6, GSAT-7, GSAT-8, GSAT 9, GSAT-10, GSAT-12, GSAT-14, GSAT-15, GSAT-16, GSAT-17, GSAT-18, GSAT-19, and GSAT-29.

4.2.4 Satellites Out of Service

- GSAT-1, GSAT-4, GSAT-5 (cancelled), GSAT-5P (launch failure), GSAT-6A satellites are presently out of services.

4.3 VSAT

4.3.1 Brief History

- The concept of the *geostationary orbit* was originated by Russian theorist *Konstantin Tsiolkovsky*, who wrote articles on space travel around the beginning of the 20th century. In the 1920s, *Hermann Oberth* and *Herman Potocnik*, described an orbit at an altitude of 35,900 kilometres (22,300 mi) whose *period* exactly matched the Earth's rotational period, making it appear to hover over a fixed point on the Earth's *equator*.

- *Arthur C. Clarke's* October 1945 *'Wireless World'* article discussed the necessary *orbital characteristics* for a *geostationary orbit* and the frequencies and power needed for communication.

- Live satellite communication was developed in the 1960s by *NASA*, which called it *Syncom 1-3*. It transmitted live coverage of the *1964 Olympics in Japan* to viewers in the *U.S.* and *Europe*. On April 6, 1965, the first commercial satellite was launched into space, *Intelsat-1*, nicknamed Early Bird.

- The first commercial Very Small Aperture Terminal (VSAT) were C band (6 GHz) receive-only systems by Equatorial Communications using *spread spectrum technology*.

- In the early 1980s, LINKABIT the predecessor to *Qualcomm* and *ViaSat*, developed the world's first Ku-band VSAT for *Schlumberger* to provide network connectivity for oil field drilling and exploration units.

- Today, the largest VSAT Ku-band network containing over 100,000 VSATs was deployed by and is operated by *Hughes Communications* for *lottery* applications.

- In 2005, WildBlue (now ViaSat) started deploying VSAT networks deploying Ka-band. ViaSat launched the highest capacity satellite ever, ViaSat-1, in 2011 to expand the *WildBlue* base under its Exede brand.

- In 2007, *Hughes Communications* started deploying Ka band VSAT sites for consumers under its *HughesNet* brand on the Spaceway 3 satellite and later in 2012 on its EchoStar XXVII/Jupiter 1 satellite. By September 2014, *Hughes* became the first Satellite Internet Provider to surpass one million active terminals.

4.3.2 Definition

- The abbreviation VSAT stands for Very Small Aperture Terminal.

- **A VSAT is a two-way satellite ground station with a dish antenna that is smaller than 3.8 meters.** It is a two way Earth station that transmits and receives data from satellite.

- A VSAT is a satellite communication system that serves home and business users.

- A VSAT is less than three meters tall and is capable of both narrow and broadband data to satellites in orbit in real-time. The data can then redirected to other remote terminals or hubs around the planet.

4.3.3 Key Features

1. VSAT is a data transmission technology.
2. VSAT can be used in place of a large physical network.
3. There can be a latency issue that wouldn't exist with a physical network.
4. Weather can adversely impact the efficacy of a VSAT network.

4.3.4 Configurations

- Most VSAT networks are configured in one of the following topologies:

1. Star topology,
2. Mesh topology,
3. Combination of both star and mesh topologies

4.3.5 Working Principle

- A VSAT is relatively simple, consisting of two primary components: *Outdoor* and *Indoor* units. The Outdoor Unit (ODU) encompasses everything you see outside the building, where the terminal is located. This is the equipment that enables the terminal to transmit and receive signals to and from the satellite. The ODU includes the following:

1. Reflector,
2. Feed,
3. Block Upconverter (BUC),
4. Low Noise Block Downconverter (LNB)

- The *Outdoor Unit* is connected to the *indoor* VSAT component with an Intra-Facility Link (IFL) cable. A coaxial cable is often used for this purpose.

- The Indoor Unit (IDU) consists of the satellite *'modem'* and an IP router which connects to an *Ethernet interface*, and it is here that the data being received by end users, and data and commands, are entered into and transmitted back to the satellite.

- VSAT networks typically follow a network architecture form called a VSAT *"Star Network"*, which consists of multiple VSAT terminals spread throughout a designated service area and controlled by a central hub computer.

- VSAT networks can also make use of a *'Mesh topology'*, a method where one terminal will transmit information to other terminals via the satellite, minimizing the need to an uplink site.
- However, Star and Mesh topologies are not mutually exclusive, and it can be more cost-effective to use them together via multiple uplink sites connected through multi-star network architecture.

4.3.6 Applications

1. In narrowband data. e.g. sale transaction.
2. Using debit cards or credit cards, RFID data.
3. In broadband data, e.g. Internet access.
4. Mobile communications.
5. Maritime communications.

4.4 WEATHER FORECASTING

4.4.1 Brief History

- Ancient weather forecasting methods usually relied on observed patterns of events, also termed *'pattern recognition'*. For example, it might be observed that if the Sunset was particularly red, the following day often brought *fair weather*. This experience accumulated over the generations to produce *weather lore*.
- It was not until the invention of the *'electric telegraph'* in 1835 that the modern age of weather forecasting began. By the late 1840s, the *telegraph* allowed reports of weather conditions from a wide area to be received almost instantaneously, allowing forecasts to be made from knowledge of weather conditions further upwind.
- *Francis Beaufort and Protege Robert FitzRoy*, Brittish Naval officers, credited to the birth of forecasting as a science. *Francis Beaufort* developed the *Wind Force Scale* (WFS) and Weather Notation coding, which he was to use in his journals for the remainder of his life. He also promoted the development of reliable *'tide tables'* around British shores, and with his friend *William Whewell*, expanded weather record-keeping at 200 British Coast guard stations.
- A storm in 1859 that caused the loss of the Royal Charter inspired *Protégé FitzRoy* to develop charts to allow predictions to be made, which he called *"forecasting the weather"*. *Protégé FitzRoy* published a book titled *"Weather Book"* in 1863, which was far in advance of the scientific opinion of the time.
- In 1922, English scientist *Lewis Fry Richardson* published *"Weather Prediction By Numerical Process"*. The first computerised weather forecast was performed by a team composed of American meteorologists *Jule Charney, Philip Thompson, Larry Gates*, and Norwegian meteorologist *Ragnar Fjortoft*, applied mathematician John von Neumann, and ENIAC programmer Klara Dan von Neumann.
- The practical use of numerical weather prediction began in 1935, spurred by the development of programmable electronic computers.

4.4.2 Definition

- **Weather forecasting is the application of science and technology to predict the conditions of the atmosphere for a given location.**
- Weather forecasts are made by collecting quantitative data about the current state of the atmosphere at a given place and using meteorology to project how the atmosphere will change.

4.4.3 Need

- The purpose of weather forecasting is to provide as accurate as possible prediction of what the weather will be like in the near future. They are important to most aspects of day to day life, including aviation, boating, other modes of transportation, farming, tourism, sports, etc. Without accurate weather forecasts people involved in activities like the ones we have listed may end up in dangerous situations they were unprepared for and end up injured or worse.
- Pilots need to know the weather to plan their flights, sailors need to know what the weather will be like to plan their activities, and farmers need to know what the weather will be like to help them plan watering, fertilizer and pesticide application, and harvest activities, to name a few. For this purpose, there is a need of weather forecasting.

4.4.4 Techniques of Forecasting

1. **Persistence:** The simplest method of forecasting the weather is *persistence* that relies upon today's conditions to forecast the conditions tomorrow. This method of forecasting strongly depends upon the presence of a stagnant weather pattern.

2. **Use of a barometer:** Measurements of barometric pressure and the pressure tendency (the change of pressure over time) have been used in forecasting, since the late 19th century.

3. **Looking at the sky:** Along with pressure tendency, the condition of the sky is one of the more important parameters used to forecast weather in mountainous areas.

4. **Nowcasting:** The forecasting of the weather within the next six hours is often referred to as nowcasting. In this time range it is possible to forecast smaller features such as individual showers and thunderstorms with reasonable accuracy, as well as other features too small to be resolved by a computer model.

5. **Use of forecast models:** In the past, the human forecaster was responsible for generating the entire weather forecast based upon available observations. Today, human input is generally confined to choosing a model biases and performance.

6. **Analog technique:** The analog technique is a complex way of making a forecast, requiring the forecaster to remember a previous weather event that is expected to be mimicked by an upcoming event.

4.4.5 Areas of Applications

- There are a number of sectors with their own specific needs for weather forecasts and specialist services are provided to these users.

1. **Air traffic:** Because the aviation industry is especially sensitive to the weather, accurate weather forecasting is essential. Fog or exceptionally low ceilings can prevent many aircraft from landing and thunderstorms are a problem for all aircraft because of severe turbulence due to their updrafts and outflow boundaries, icing due to the heavy precipitation, as well as large hail, strong winds, and lightning, all of which can cause severe damage to an aircraft in flight. Volcanic ash is also a significant problem for aviation, as aircraft can lose engine power within ash clouds.

2. **Marine:** Commercial and recreational use of waterways can be limited significantly by wind direction and speed, wave periodicity and heights, tides, and precipitation. These factors can each influence the safety of marine transit.

3. **Agriculture:** Farmers rely on weather forecasts to decide what work to do on any particular day. For example, drying hay is only feasible in dry weather. Prolonged periods of dryness can ruin cotton, wheat, and corn crops. While corn crops can be ruined by drought, their dried remains can be used as a cattle feed substitute in the form of silage. Frosts and freezes play havoc with crops both during the spring and fall.

4. **Forecasting:** Weather forecasting of wind, precipitations and humidity is essential for preventing and controlling wildfires. Different indices, like the forest fire weather index and the Haines Index, have been developed to predict the areas more at risk to experience fire from natural or human causes. Conditions for the development of harmful insects can be predicted by forecasting the evolution of weather, too.

5. **Utility companies:** Electricity and gas companies rely on weather forecasts to anticipate demand, which can be strongly affected by the weather. Similarly, in summer a surge in demand can be linked with the increased use of air conditioning systems in hot weather.

4.5 REMOTE SENSING

4.5.1 Brief History

- The modern discipline of remote sensing arose with the development of flight. The balloonist *G. Tournachon* made photographs of Paris from his balloon in 1858. Messenger pigeons, kites, rockets and unmanned balloons were also used for early images.

- Systematic *aerial photography* was developed for military surveillance and reconnaissance purposes beginning in *World War I* and reaching a climax during the *Cold War* with the use of modified combat aircraft or specifically designed collection platforms.
- A more recent development is that of increasingly smaller *sensor pods* such as those used by law enforcement and the military, in both manned and unmanned platforms.
- The development of *artificial satellites* in the latter half of the 20th century allowed remote sensing to progress to a global scale as at the end of the Cold War. Instrumentation aboard various Earth observing and weather satellites such as Landsat, the Nimbus and more recent missions such as RADARSAT and UARS provided global measurements of various data for civil, research, and military purposes. Space probes to other planets have also provided the opportunity to conduct remote sensing studies in extraterrestrial environments, synthetic aperture radar aboard the Magellan spacecraft provided detailed topographic maps of Venus, while instruments aboard SOHO allowed studies to be performed on the Sun and the solar wind.
- Recent developments include, beginning in the 1960s and 1970s with the development of *image processing* of satellite imagery. Several research groups in Silicon Valley including NASA Ames Research Center, GTE, and ESL Inc. developed Fourier transform techniques leading to the first notable enhancement of imagery data. In 1999 the first commercial satellite (IKONOS) collecting very high resolution imagery was launched.

4.5.2 Definition

- **Remote sensing is the acquisition of information about an object or phenomenon without making physical contact with the object and thus in contrast to on-site observation, especially the Earth.**
- In current usage, the term "remote sensing" generally refers to the use of satellite- or aircraft-based sensor technologies to detect and classify objects on Earth.
- **Remote sensing is a process of detecting and monitoring the physical characteristics of an area by measuring its reflected and emitted radiation at a distance typically from space craft or satellite.**
- Remote sensing is the science of obtaining information about objects or areas from a distance, typically from aircraft or satellites.

4.5.3 Types

- Remote sensing may be split into two types as under:
  1. Passive Remote Sensing
  2. Active Remote Sensing

1. **Passive Remote Sensing:** Passive sensors gather radiation that is emitted or reflected by the object or surrounding areas. Reflected sunlight is the most common source of radiation measured by passive sensors. Examples of passive remote sensors include film photography, infrared, charge-coupled devices, and radiometers.
2. **Active Remote Sensing:** Active collection, on the other hand, emits energy in order to scan objects and areas whereupon a sensor then detects and measures the radiation that is reflected or backscattered from the target. RADAR and LiDAR are examples of active remote sensing where the time delay between emission and return is measured, establishing the location, speed and direction of an object.

4.5.4 Description

- Remote sensing makes it possible to collect data of dangerous or inaccessible areas. Military collection during the **Cold War** made use of stand-off collection of data about dangerous border areas. Remote sensing also replaces costly and slow data collection on the ground, ensuring in the process that areas or objects are not disturbed.
- Orbital platforms collect and transmit data from different parts of the **electromagnetic spectrum**, which in conjunction with larger scale aerial or ground-based sensing and analysis, provides researchers with enough information to monitor trends such as **El Nine** and other natural long and short term phenomena.

4.5.5 Application Areas

1.	Geography	2.	Land surveying
3.	Hydrology	4.	Ecology
5.	Meteorology	6.	Oceanography
7.	Glaciology	8.	Geology
9.	Agricultural	10.	Natural resource management
11.	Climatology	12.	Mineralogy
13.	Biology and defence	14.	Environmental measurements

4.5.6 Applications

1.	Aerial traffic control	2.	Early warning
3.	Meteorological data	4.	Monitoring of speed limits
5.	Wind speed and direction	6.	Digital elevation models
7.	Wide range and data collection	8.	Mapping on the seafloor
9.	Surface ocean currents and directions	10.	Sea level, tides and wave direction

11. Weapon ranging

12. Concentration of various chemicals in the atmosphere

13.	Heights of objects	14.	Features on the ground
15.	Collecting reflected and emitted radiation	16.	Detect the emission spectra
17.	Damage to infrastructure from war or disasters	18.	Target tracking
19.	Topographic maps	20.	Modelling terrestrial habitat features
21.	Prospect for minerals	22.	Detect or monitor land usage

23. Examine the health of indigenous plants and crops

24. Examine the health of farming regions or forests

25.	Indicate water quality parameters	26.	Position and deformation imaging

27. Detecting, ranging and measurements of underwater objects and terrain.

28.	Locate and measure Earthquakes	29.	Detecting water waves and level

4.6 LANDSAT PROGRAM

4.6.1 Brief History

- The Hughes Aircraft company's *Santa Barbara Research Center* initiated, designed, and fabricated the first three Multispectral Scanners (MSS) in 1969. The first prototype MSS was completed within nine months, in the fall of 1970.

- Working at NASA's Goddard Space Flight Center, *Valerie L. Thomas* managed the development of early Landsat image processing software systems. He showed for the first time that global crop monitoring could be done with Landsat satellite imagery.

- The program was initially called the Earth Resources Technology Satellites Program (ERTSP), which was used from 1966 to 1975. In 1975, the name was changed to LANDSAT. In 1979, President of the U.S. *Jimmy Carter's* Presidential Directive transferred LANDSAT operations from NASA to NOAA (National Oceanic and Atmospheric Administration).

- In 1985 when the Earth Observation Satellite Company (EOSAT), a partnership of *Hughes Aircraft* and RCA, was selected by NOAA to operate the LANDSAT system with a ten-year contract. EOSAT operated LANDSAT 4 and LANDSAT 5, had exclusive rights to market LANDSAT data, and was to build LANDSATs 6 and 7.

- In 1989, NOAA directed that LANDSATs 4 and 5 be shut down due to finding problem. By the end of 1992, EOSAT ceased processing LANDSAT data. But LANDSAT 6 was finally launched on October 5, 1993, but was lost in a launch failure.

- Processing of LANDSATs 4 and 5 data was resumed by EOSAT in 1994. NASA finally launched LANDSAT 7 on April 15, 1999. On July 23, 1972 the ERTS was launched.

Table 4.1

Instrument	Launched	Terminated	Duration
LANDSAT 1	July 23, 1972	January 6, 1978	5 years, 6 months and 14 days
LANDSAT 2	January 22, 1975	February 25, 1982	7 years, 1 month and 3 days
LANDSAT 3	March 5, 1978	March 31, 1983	5 years and 26 days
LANDSAT 4	July 16, 1982	December 14, 1993	11 years, 4 months and 28 days
LANDSAT 5	March 1, 1984	June 5, 2013	29 years, 3 months and 4 days
LANDSAT 6	October 5, 1993	October 5, 1993	0 days
LANDSAT 7	April 15, 1999	Still active	20 years, 9 months and 3 days
LANDSAT 8	February 11, 2013	Still active	6 years, 11 months and 7 days
LANDSAT 9	December 2020 (expected)	-	-

- The most recent, LANDSAT 8, was launched on February 11, 2013. The instruments on the LANDSAT satellites have acquired millions of images. The images, archived in the United States and at LANDSAT receiving stations around the world,

4.6.2 Definition

- The LANDSAT program is the longest-running enterprise for acquisition of satellite imagery of Earth.
- **Any of various satellite used together data for images of the Earth's surface and coastal region is called a LANDSAT.**
- LANDSATs are equipped with sensors that respond to Earth-reflected sunlight and infrared radiation.
- LANDSAT sensors record reflected and emitted energy from Earth in various wavelengths of the electromagnetic spectrum. It is the digital information that makes remotely sensed data invaluable.

4.6.3 Application Areas

1. Agriculture
2. Forestry
3. Geology
4. Hydrology
5. Cartography
6. Environment
7. Climatology
8. Oceanography
9. Meteorology
10. Agro Industry

4.6.4 Applications

1. Understand seeded crops and fight insurance found.
2. Access the severity of wildfires.
3. Track deforestation in rainforests.
4. Monitor irrigation of crops.
5. Watch the growth of cities world wide.
6. Find high yield fishery areas.
7. Generate land cover map of man grove forest.
8. Quantity the amount of water loss and changes to show line.
9. Area estimation.
10. Display activeness for timber production.
11. Generate climate change knowledge.
12. Find areas for conservation for forests.

4.7 SATELLITE NAVIGATION

4.7.1 Brief History

- Ground based radio navigation has long been practiced. The DECCA, LORAN, GEE and Omega systems used terrestrial longwave radio transmitters which broadcast a radio pulse from a known "*master*" location, followed by a pulse repeated from a number of "slave" stations. The delay between the reception of the master signal and the slave signals allowed the receiver to deduce the distance to each of the '*slaves*', providing a fix.

- The first satellite navigation system was Transit, a system deployed by the US military in the 1960s. Transit's operation was based on the *Doppler effect*: the satellites travelled on well-known paths and broadcast their signals on a well-known radio frequency. The received frequency will differ slightly from the broadcast frequency because of the movement of the satellite with respect to the receiver.

- By monitoring this '*frequency shift*' over a short time interval, the receiver can determine its location to one side or the other of the satellite, and several such measurements combined with a precise knowledge of the satellite's orbit can fix a particular position. Satellite orbital position errors are induced by variations in the gravity field and radar refraction, among others. These were resolved by a team led by *Harold L Jury* of Pan Am Aerospace Division in Florida from 1970-1973. Using real-time data assimilation and recursive estimation, the systematic and residual errors were narrowed down to a manageable level to permit accurate navigation.

- As a satellite's orbit deviated, the USNO would send the updated information to the satellite. Subsequent broadcasts from an updated satellite would contain its most recent ephemeris. The modern systems are more direct.

- Beeping radio signals for '*Sputnik*' inspired idea of using satellite to navigate. The idea for the first space-based navigation system was born at the Johns Hopkins University's Applied Physics Laboratory (APL) in 1957, as scientists listened to the radio signals from '*Sputnik*', the world's first satellite. By 1964, the Navy was using radio signals from its own satellites to navigate submarines and surface ships, a system they named Transit.

4.7.2 Definition

- **A satellite navigation or satnav system is a system that uses satellites to provide autonomous geo-spatial positioning.**

- Satnav systems operate independently of any telephonic or internet reception.

- **Satellite navigation service is an emerging satellite based system with commercial and strategic applications.**

4.7.3 Global Navigation Satellite Systems

- The Global Navigation Satellite Systems (GNSSs) are as under:
    1. GPS (Global Positioning System)
    2. GLONASS (Global Navigation Satellite System)
    3. Galileo
    4. BeiDou

4.7.4 Regional Navigation Satellite Systems (RNSS)

- **The Navigation with Indian Constellation or NAVIC is an autonomous regional satellite navigation system developed by *Indian Space Research Organization* (ISRO), which would be under the total control of *Indian government*.**

- The government approved the project in May 2006, with the intention of the system completed and implemented on 28 April 2016. It will consist of a constellation of 7 navigational satellites. Three of the satellites will be placed in the Geostationary Orbit (GEO) and the remaining 4 in the Geosynchronous orbit (GSO) to have a larger signal footprint and lower number of satellites to map the region. It is intended to provide an all-weather absolute position accuracy of better than 7.6 meters throughout India and within a region extending approximately 1,500 km around it. A goal of complete Indian control has been stated, with the space segment, ground segment and user receivers all being built in India.

- All 7 satellites, IRNSS-1A, IRNSS-1B, IRNSS-1C, IRNSS-1D, IRNSS-1E, IRNSS-1F, and IRNSS-1G, of the proposed constellation were precisely launched on 1 July 2013, 4 April 2014, 16 October 2014, 28 March 2015, 20 January 2016, 10 March 2016 and 28 April 2016 respectively from Satish Dhawan Space Centre.

- It covers India and a region extending 1,500 km (930 mi) around it, with plans for further extension. An Extended Service Area lies between the primary service area and a rectangular area 1,500–6,000 km beyond borders. The system at present consists of a constellation of seven satellites, with two additional satellites on ground as stand-by.

- The constellation was in orbit as of 2018, and the system was operational from early 2018 after a system check. NAVIC provides two levels of service, the *"standard positioning service"*, which will be open for civilian use, and a *"restricted service"* for authorized users including military.

- There are plans to expand NAVIC system by increasing constellation size from 7 to 11.

4.7.5 Working Principle

- A satellite navigation system is also known as a *'satnav'* system. A satnav system receiver can be used to locate latitude, altitude, velocity and time information. The operation of the satellite navigation is based on the *Dopper effect*.

- The satellites broadcast a signal that contains orbital data and the exact time the signal is transmitted. The orbital data is transmitted in a data message that is superimposed on a code that serves as a timing reference.

- The satellite uses an *'atomic clock'* (most accurate time and frequency standards known) to maintain synchronization of all the satellites in the constellation. The receiver compares the time of broadcast encoded in the transmission with the time of reception measured by an internal clock, thereby measuring the time-of-flight to the satellite.

- The receiver measures signals from several satellites at the same time so that it can use triangulation to determine its location. Triangulation is the process of determining the location of a point by measuring the angles to it from two known points.

- The precise satellite locations are included in the transmission and the time-of-flight of the signal is used to calculate the distance to each satellite. The receiver then does some math and calculates its location on the Earth. The more satellites the receiver can track, the more accurate the location calculation.

- The receiver calculates 4 parameters; latitude, longitude, altitude and time. As a result, the receiver generally needs to see at least 4 satellites to calculate the 4 unknowns. It can give estimates for the values with fewer satellites, but the potential error increases.

- The triangulation math is not that complicated, but the fact that the known points, the satellites, are moving very fast and the fact that the Earth is a curved surface adds quite a bit of complexity. In addition, the Earth is not a perfect sphere and is not uniformly shaped or curved. This adds some error depending on how far off the average curvature a specific location is. For this reason, local augmentation systems are used. The receiver can also use regional data sets that better describe the local geography and ultimately give a more accurate position.

4.7.6 Applications

1. Delivery of weapons to targets.
2. Military forces to be directed and to locate themselves more easily.
3. Determine location of users and the location of other people or objects.
4. Public and private sectors across numerous market segments such as science, transport, agriculture etc.

4.8 MOBILE SATELLITE SERVICES (MSS)

4.8.1 Definition

- **Mobile satellite services (MSS) refers to networks of communications satellites intended for use with mobile and portable wireless telephones.**

- **Mobile satellite services (MSS) is a satellite-based communication system that offers the services commonly available in a terrestrial mobile networks.**

4.8.2 Major Types

- There are three major types of MSS as under:

  1. Aeronautical MSS (AMSS),

  2. Land MSS (LMSS), and

  3. Maritime MSS (MMSS).

4.8.3 Working Principle

- A telephone connection using MSS in similar to a cellular telephone link, except the repeaters are in orbit around the Earth, rather than on the surface. MSS repeaters can be placed on geostationary, Medium Earth Orbit (MEO) or Low Earth Orbit (LEO) satellites.

- Provided there are enough satellites in the system, and provided they are properly spaced around the globe, an MSS can link any two wireless telephone sets at any time, no matter where in the world they are located. MSS systems are interconnected with land-based cellular networks.

- As an example of how MSS can work, consider telephones in commercial airliners. These sets usually link into the standard cellular system. This allows communication as long as the aircraft is on a line of sight with at least one land-based cellular repeater. Coverage is essentially continuous over most developed countries. But coverage is spotty over less well-developed regions, and is nonexistent at most points over the oceans. Using an MSS network, the aircraft can establish a connection from any location, no matter how remote.

4.8.4 Identified Services

- An S-band mobile satellite service (MSS) was added to INSAT system with the launch of INSAT-36 in 2002 and GSAT in 2003. The following two classes of services were identified for MSS:

  1. A small portable satellite terminal that works with INSAT for voice/data communication has been developed with the participation of Indian industries. The terminal is useful for voice communication especially during disasters when other means of communication break down. It can be used from any location in India for emergency communication. Transmit and receive frequencies of the terminal are in S-Band.

  2. The portable terminal is connected to the EPABX at central hub station through satellite channel and hence could be considered as an extension of EPABX and call could be made between any satellite terminals and local phones on EPABX. Central hub station is located at SAC, Ahmedabad.

- The report forecasts the market sizes and trends for MSS in the following sub-marketing:

  1. On the basis of services:

     (a) Voice services,   (b) Video services,

     (c) Data services,   (d) Tracking and Monitoring services

  2. On the basic of access type:

     (a) Land mobile   (b) Maritime

     (c) Aeronautical

3. On the basic of industry:

    (a) Oil and gas   (b) Media and entertainment

    (c) Mining   (d) Military and deference

    (e) Aviation   (f) Transportation

4. On the basic of region:

    (a) North America (NA)   (b) Europe

    (c) Asian Pacific (APAC)   (d) Middle East and Africa (MEA)

    (e) Latin America (LA)

4.8.5 Applications

1. Application development and platforms

2. Application Architecture and Platforms

3. Application and Product Porfolio Governance

4. Application Leaders

5. Application Architecture and Platform for Technical Professionals

6. Customer Experience Leaders

7. Development and Integration Professionals

Practice Questions

1. What is INTELSAT ?

2. Describe brief history of INTELSAT.

3. What is INSAT ?

4. List the satellite INSAT in service.

5. What is INSAT system ?

6. What is VSAT ? State its applications.

7. State key features of VSAT.

8. What are the configurations of VSAT ?

9. Explain working principle of VSAT.

10. What is weather forecasting ?

11. Why do you need weather forecasting ?

12. What is the purpose of weather forecasting ?

13. List the various techniques of weather forecasting.

14. State important applications of weather forecasting.

15. What is remote sensing ?

16. What are active and passive remote sensing ?

17. State important applications of remote sensing.

18. What is LANDSAT program ?

19. State important applications of LANDSAT program.

20. What is satellite navigation ?

21. What is regional navigation satellite system ?

22. Explain the working principle of satellite navigation.

23. State important applications of satellite navigation.

24. What is mobile satellite service ?

25. Explain the working principle of mobile satellite services (MSS).

26. What are identified services provided by MSS ?

27. What are important applications of MSS ?

LASER AND SATELLITE COMMUNICATION

Laser & Satellite Communication link analysis, Optical satellite link, Tx & Rx, Satellite, Beam acquisition, Tracking & pointing, Cable channel frequency, Head end equation, Distribution of signal, Network specifications and architecture, Optical fibre CATV system.

5.1 LASER

5.1.1 Brief History

- In 1917, Albert Einstein established the theoretical foundations for the Laser and the Maser in the paper *Zur Quantentheorie der Strahlung* (On the Quantum Theory of Radiation). In 1928, *Rudolf W. Ladenburg* confirmed the existence of the phenomena of *stimulated emission* and *negative absorption*.

- In 1939, *Valentin A. Fabrikant* predicted the use of stimulated emission to amplify "short" waves. In 1947, *Willis E. Lamb* and *R.C. Retherford* found apparent stimulated emission in hydrogen spectra and effected the first demonstration of *stimulated emission*. In 1950, *Alfred Kastler* proposed the method of optical pumping, experimentally confirmed, two years later, by *Brossel, Kastler*, and *Winter*. In 1966, *Alfred Kastler* received a 'Nobel Prize' for Physics.

- In 1957, *Charles Hard Townes* and *Arthur Leonard Schawlow*, then at Bell Laboratories, began a serious study of the infrared (IR) laser. As ideas developed, they abandoned infrared radiation to instead concentrate upon visible light.

- The concept originally was called an "*optical Maser*". In 1958, Bell Labs filed a patent application for their proposed optical Maser; and *Schawlow* and *Townes* submitted a manuscript of their theoretical calculations to the *Physical Review*.

- Simultaneously, at Columbia University, graduate student *Gordon Gould* was working on a doctoral thesis about the energy levels of excited thallium. When *Gould* and *Townes* met, they spoke of radiation emission, as a general subject; afterwards, in November 1957, *Gould* noted his ideas for a "laser", including using an open resonator (later an essential Laser-device component). Moreover, in 1958, *Prokhorov* independently proposed using an open resonator, the first published appearance (in the USSR) of this idea. Elsewhere, in the U.S., *Schawlow* and *Townes* had agreed to an open-resonator laser design – apparently unaware of *Prokhorov's* publications and *Gould's* unpublished laser work.

- At a conference in 1959, *Gordon Gould* published the term LASER in the paper *"The LASER"*. On May 16, 1960, Theodore H. Maiman operated the first functioning laser at Hughes Research Laboratories, Malibu, California, ahead of several research teams, including those of *Townes*, at Columbia University, *Arthur Schawlow* and *Gordon Gould* working at different places.

- *Theodore Maiman's* functional laser used a flashlamp-pumped synthetic ruby crystal to produce red laser light at 694 nanometers wavelength.

- Iranian physicists *Ali Javan*, and *William R. Bennett*, and *Donald Herriott*, constructed the first gas laser, using helium and neon that was capable of continuous operation in the infrared.

- Later, *Javan* received the *Albert Einstein* Award in 1993. *Basov* and *Javan* proposed the semiconductor laser diode concept.

- In 1962, *Robert N. Hall* demonstrated the first *Laser diode* device, which was made of gallium arsenide and emitted in the near-infrared band of the spectrum at 850 nm. Later that year, *Nick Holonyak, Jr.* demonstrated the first semiconductor laser with a visible emission.

- In 2015, researchers made a white Laser, whose light is modulated by a synthetic nanosheet made out of zinc, cadmium, sulfur, and selenium that can emit red, green, and blue light in varying proportions, with each wavelength spanning 191 nm.

- In 2017, researchers at TU Delft demonstrated an AC Josephson junction microwave laser. Since the laser operates in the superconducting regime, it is more stable than other semiconductor-based lasers. The device has potential for applications in quantum computing. In 2017, researchers at TU Munich demonstrated the smallest mode locking laser capable of emitting pairs of phase-locked picosecond laser pulses with a repetition frequency upto 200 GHz.

- In 2017, researchers from the Physikalisch-Technische Bundesanstalt (PTB), together with US researchers from JILA, a joint institute of the National Institute of Standards and Technology (NIST) and the University of Colorado Boulder, established a new world record by developing an erbium-doped fiber Laser with a linewidth of only 10 mHz.

5.1.2 Definition

- The term "Laser" Originated as an acronym for "**Light Amplification by Stimulated Emission of Radiation**".

- A **Laser** is a device that emits light through a process of optical_amplification based on the stimulated emission of electromagnetic_radiation.

- Most Lasers emit nearly monochromatic light with a narrow frequency spectrum. The first laser was built in 1960 by Theodore H. Maiman at Hughes Research Laboratories, based on theoretical work by Charles Hard Townes and Arthur Leonard Schawlow.

- A **Laser** is a machine that makes an amplified, single-colour source of light. It uses special gases or crystals to make the light with only a single colour. Laser is a device that produces a nearly parallel, nearly monochromatic, and coherent beam of light by exciting atoms to a higher energy level and causing them to radiate their energy in phase.

5.1.3 Mechanism of Working

- A Laser creates light by special actions involving a material called an *"optical gain medium"*. Energy is put into this material using an *'energy pump'*. This can be electricity, another light source, or some other source of energy. The energy makes the material go into what is called an excited state.

- This means the electrons in the material have extra energy, and after a bit of time they will lose that energy. When they lose the energy they will release a photon (a particle of light). The type of optical gain medium used will change what colour (wavelength) will be produced. Releasing photons is the *"stimulated emission of radiation"* part of laser.

- Many things can radiate light, like a light bulb, but the light will not be organized in one direction and phase. By using an electric field to control how the light is created, this light will now be one kind, going in one direction. This is *"coherent radiation"*.

- At this point, the light is still weak. The mirrors on either side bounce the light back and forth, and this hits other parts of the optical gain medium, causing those parts to also release photons, generating more light (*"light amplification"*). When all of the optical gain medium is producing light, this is called saturation and creates a very strong beam of light at a very narrow wavelength, which we would call a laser beam.

5.1.4 Principle of Working

- In Lasers, photons are interacted in the following three ways with the atoms:
  1. Absorption of radiation
  2. Spontaneous emission
  3. Stimulated emission
- The three different mechanisms are shown below (Fig. 5.1):
  1. **Absorption:** An atom in a lower level absorbs a photon of frequency hv and moves to an upper level.

  2. **Spontaneous emission:** An atom in an upper level can decay spontaneously to the lower level and emit a photon of frequency hv if the transition between E_2 and E_1 is radiative. This photon has a random direction and phase.

  3. **Stimulated emission:** An incident photon causes an upper level atom to decay, emitting a "stimulated" photon whose properties are identical to those of the incident photon. The term "stimulated" underlines the fact that this kind of radiation only occurs if an incident photon is present. The amplification arises due to the similarities between the incident and emitted photons.

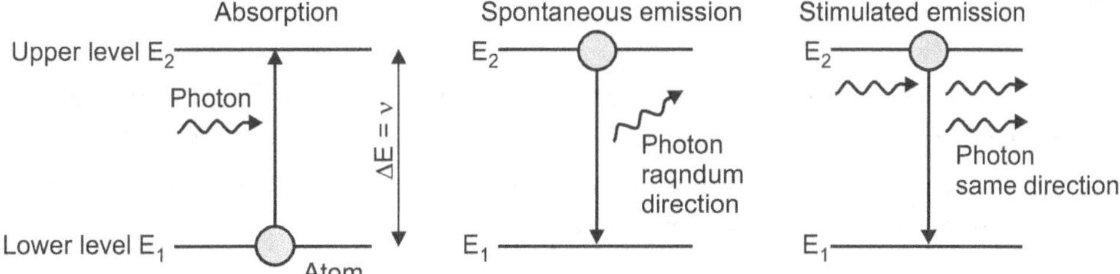

Fig. 5.1 : Mechanism of the interaction between an atom and a photon

(The photon has an energy hv equal to the difference between the two atomic energy levels)

4. **Optical Pumping:**

(i) **Definition :**

- **Optical pumping** is a process in which light is used to raise (or "pump") electrons from a lower energy level in an atom or molecule.
- Optical pumping technique was first developed by Alfred Kastler in the early 1950s and is primarily **used to pump laser medium during laser construction.** It is also cyclically pump electrons bound within an atom or molecule to a well-defined quantum state. It is commonly used in laser construction, to pump the active laser medium so as to achieve population inversion.

(ii) **Working:**

- When a system consisting of atoms with random orientation of magnetic fields of atoms will be realized towards the direction of light. This results in the rearrangement of magnetic energy levels. In some cases, a group of atoms will be oriented to create a magnetic field within the system.
- Optical pumping often takes place in laser action which involves absorption of photons of light to raise the energy levels of electrons. These electrons remain at higher energy state until they are activated to release their stored energy in the form of a laser beam. Several types of lasers can be optically pumped. The most common optically pumped lasers are doped-insulator solid-state lasers consisting of glass, ceramic or laser crystal host medium.

(iii) **Optical Pump Sources:**

  1. Laser diodes
  2. Discharge lamps
  3. Dye lasers
  4. Titanium-sapphire lasers

(iv) Applications:

(a) It is used to enhance the nuclear spin polarization.

(b) It is used for improvements in NMR detection sensitivity.

(c) It is used for development of optically enhanced NMR and MRI.

(d) It is used for potential NMR applications include void-space imaging of materials and living organisms; probing structure and dynamics in proteins and inclusion complexes; enhanced NMR of bulk and nanostructured materials; quantum computing; fundamental studies of various photophysical and photochemical processes; and low-field and remotely-detected NMR and MRI.

(e) It is used for optical nuclear polarization (ONP) in organic molecular crystals.

5. Population Inversion:

(i) Definition:

- Population inversion is the redistribution of atomic energy levels that takes place in a system so that laser action can occur.

- **Population inversion is a state of medium where a higher-lying electronic level has a higher population than a lower-lying level.**

(ii) Need:

- Population inversion is required in laser so that stimulated emission is more probable than induced absorption. Generally, more number of atoms are in lower energy states. When a photon having energy equal to the energy difference of two states of atom is incident, the probability of getting absorbed is one as lower energy atoms are more.

- However, in popular inversion state, the probability of photon causing stimulated inversion is more, as more atoms are in higher energy state causing amplification of light. Therefore, there is a heard of popular inversion in laser.

(iii)

- A population inversion is required for Laser operation, but cannot be achieved in theoretical group of atoms with two energy levels, when they are in thermal equilibrium. In fact, any method by which the atoms are directly and continuously excited from the ground state to the excited state (such as optical absorption) will eventually reach equilibrium with the de-exciting processes of spontaneous and stimulated emission.

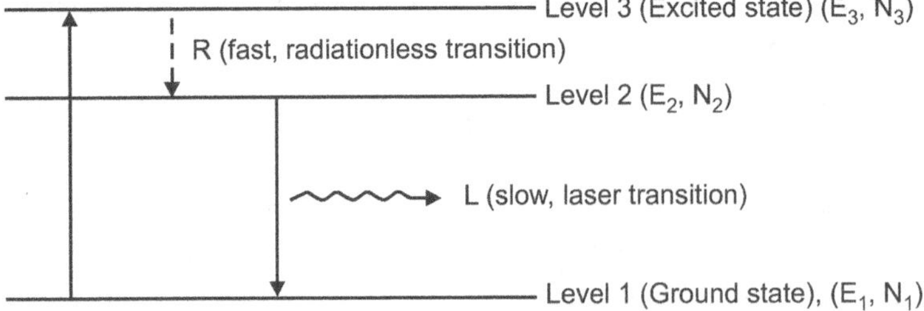

Fig. 5.2 : A three-level laser energy diagram

- To achieve non-equilibrium conditions, an indirect method of populating the excited state must be used. To understand how this is done, we may use a slightly more realistic model, that of a three-level laser. Again consider a group of N-atoms, this time with each atom able to exist in any of three energy states, levels 1, 2 and 3, with energies E_1, E_2 and E_3, and populations N_1, N_2 and N_3, respectively. Also we assume that $E_1 < E_2 < E_3$; that is, the energy of level 2 lies between that of the ground state and level 3.

- Initially, the system of atoms is at thermal equilibrium, and the majority of the atoms will be in the ground state, i.e., $N_1 \approx N$, $N_2 \approx N_3 \approx 0$. If we now subject the atoms to light of a frequency

$v_{13} = \dfrac{1}{h}(E_3 - E_1)$, the process of optical absorption will excite electrons from the ground state to level 3. This process is called pumping, and does not necessarily always directly involve light absorption; other methods of exciting the laser medium, such as electrical discharge or chemical reactions, may be used.

- The level 3 is sometimes referred to as the pump level or pump band, and the energy transition $E_1 \rightarrow E_3$ as the pump transition, which is shown as the arrow marked P in Fig. 5.2.

- Upon pumping the medium, an appreciable number of atoms will transition to level 3, such that $N_3 > 0$. To have a medium suitable for laser operation, it is necessary that these excited atoms quickly decay to level 2. The energy released in this transition may be emitted as a photon (spontaneous emission), however in practice the $E_3 \rightarrow E_2$ transition labelled **R** in Fig. 5.2 is usually radiationless, with the energy being transferred to vibrational motion (heat) of the host material surrounding the atoms, without the generation of a photon.

- An electron in level 2 may decay by spontaneous emission to the ground state, releasing a photon of frequency v_{12} (given by $E_2 - E_1 = hv_{12}$), which is shown as the transition **L**, called the laser transition in Fig. 5.2.

- If the lifetime of this transition, τ_{21} is much longer than the lifetime of the radiationless $3 \rightarrow 2$ transition τ_{32} (if $\tau_{21} \gg \tau_{32}$, known as a favourable lifetime ratio), the population of the E_3 will be essentially zero ($N_3 \approx 0$) and a population of excited state atoms will accumulate in level 2 ($N_2 > 0$).

- If over half the N atoms can be accumulated in this state, this will exceed the population of the ground state N_1. A population inversion ($N_2 > N_1$) has thus been achieved between level 1 and 2, and optical amplification at the frequency v_{21} can be obtained.

- Because at least half the population of atoms must be excited from the ground state to obtain a population inversion, the laser medium must be very strongly pumped.

- This makes three-level lasers rather inefficient. In practices three-level lasers are most useful.

5.1.5 Types of Lasers

1. Semiconductor lasers
2. Solid-state lasers
3. Fiber Lasers
4. Gas lasers
5. Free electron Lasers
6. X-ray lasers
7. Chemical lasers
8. Nuclear lasers
9. Photonic crystal
10. Dye Lasers
11. Lasers

5.1.6 Applications

1. CO_2 lasers are widely used as cutting tools in industries.
2. It is routinely used by doctors on patient's bodies for blasting cancer toumers and eye surgery.
3. It is widely used for barcode scanner in departmental stores.
4. Semiconductor Laser is used in CD and DVD.
5. Fiber-optic laser is used in photonics for communication e.g. Internet.
6. It is used mainly in military for Laser guided weapons and missiles.
7. The space-based X-rays are used to destroy incoming energy missiles.
8. Solid state lasers are used to damage destroy energy equipment.
9. It is used in laser disc player and laser printers.
10. It is used in fiber-optic communication, free-space optical communication and laser communication in space.

11. It is used in industries for conversing thin materials, welding, material heat treatment, making parts, engraving and banding.

12. It is used in industries for additive manufacturing or 3D prioting processes such as selective laser interring, and selective laser melting, non-contact measuring of parts and 3D scanning, laser cleaning.

13. It is used in military for marking targets, guiding functions, missile defence, electro optical counter measure (EOCM), LIDAR blinding treeps.

14. It is used in low enforcement for LiDAR, traffic, enforcement, and also for Latent finger detection, in the forensic identification field.

15. It is used in research for spectroscopy, laser ablation, laser annealing, Laser Scattering, Laser interferometry, LiDAR, Laser capture microdisection, fluorescence microscopy, metrology, Laser cooling.

16. It is used in consumer products for laser, printers, barcode scanners, thermometers, Laser pointers, holograms, bubblegrams, etc.

17. It is used for entertainment in optical disc, laser lighting displays, laser turntables, etc.

18. It is used in medicine, including laser surgery (particularly eye surgery), laser healing, kidney stone treatment, ophthalmoscope, and cosmetic skin treatment, such as ace treatment, cellulite and striae reduction, and hair removal.

19. It is used to treat cancer by shrinking or destroying tumors or precancerous growth. It is also used to treat superficial cancers that are on the surface of the body or lining of internal organs.

20. It is used to treat basic cell skin cancer and the very early stages of some cancers, such as cervical, penile, vaginal, vulvar, and non-small cell lung cancer.

21. Laser therapy is often combined with other treatments, such as surgery, chemotherapy, or radiation therapy. Laser-induced interstitial thermotherapy, or interstitial laser photocoagulation.

5.2 LINK ANALYSIS

5.2.1 Brief History

- Klerks categorized link analysis tools into 3 generations. The first generation was introduced in 1975 as the Anacpapa Chart of Harper and Harris. This method requires that a domain expert review data files, identify associations by constructing an association matrix, create a link chart for visualization and finally analyze the network chart to identify patterns of interest.

- This method requires extensive domain knowledge and is extremely time-consuming when reviewing vast amounts of data. In addition to the association matrix, the activities matrix can be used to produce actionable information, which has practical value and use to law-enforcement.

- The activities matrix, as the term might imply, centers on the actions and activities of people with respect to locations. Whereas the association matrix focuses on the relationships between people, organizations, and/or properties. The distinction between these two types of matrices, while minor, is nonetheless significant in terms of the output of the analysis completed or rendered.

- Second generation (2G) tools consist of automatic graphics-based analysis tools such as IBM i2 Analyst's Notebook, Netmap, ClueMaker and Watson. These tools offer the ability to automate the construction and updates of the link chart once an association matrix is manually created, however, analysis of the resulting charts and graphs still requires an expert with extensive domain knowledge.

- The third generation (3G) of link-analysis tools like DataWalk allow the automatic visualization of linkages between elements in a data set, that can then serve as the canvas for further exploration or manual updates.

5.2.2 Definition

- **In network theory, link analysis is a data-analysis technique used to evaluate relationships (connections) between nodes.**
- Relationships may be identified among various types of nodes (objects), including organizations, people and transactions.
- **Link analysis is a kind of knowledge discovery that can be used to visualize data to allow for better analysis, especially in the context of links, whether web links or relationship links between people or between different entities.**

5.2.3 Purposes/Need

1. Find matches in data for known patterns of interest between linked objects.
2. Find anomalies by detecting violated known as patterns.
3. Find new patterns of interest for example, in social networking, marketing on business.
- For these purposes, there is a need of link analysis in network analysis.

5.2.4 Explanation

- Link analysis is literally about analyzing the links between objects, whether they are physical, digital or relational. This requires diligent data gathering. For example, in the case of a website where all of the links and backlinks that are present must be analyzed, a tool has to shift through all of the HTML codes and various scripts in the page and then follow all the links it finds in order to determine what sort of links are present and whether they are active or dead.
- This information can be very important for search engine optimization, as it allows the analyst to determine whether the search engine is actually able to find and index the website.
- In networking, link analysis may involve determining the integrity of the connection between each network node by analyzing the data that passes through the physical or virtual links. With the data, analysts can find bottlenecks and possible fault areas and are able to patch them up more quickly or even help with network optimization.

5.2.5 Conducting Process

1. Link analysis process should have the following work flow.
2. Create entities of interest.
3. Document information in each entity.
4. Document relationships between entities.
5. Document links from.
6. Organize as you go!!

5.2.6 Knowledge Discovery

- Knowledge discovery is an iterative and interactive process used to identify, analyze and visualize patterns in data.
- Network analysis, link analysis and social network analysis are all methods of knowledge discovery, each a corresponding subset of the prior method. Most knowledge discovery methods follow these steps at the highest level:

1. Data processing
2. Transformation
3. Analysis
4. Visualization

5.2.7 Issues

1. Information overload
2. Prosecution
3. Crime prevention

5.2.8 Proposed Solution Categories

1. Heuristic-based solutions
2. Template-based solutions
3. Similarity-based solutions
4. Statistical solutions

5.2.9 Applications

1. FBI Violent Criminal Apprehension Program (ViCAP)
2. Low State Sex Crimes Analysis System
3. Minnesota State Sex Crimes Analysis System (MIN/SCAP)
4. Washington State Homicide Investigation Tracking System (HITS)
5. New York State Homicide Investigation & Lead Tracking (HALT)
6. New Jersey Homicide Evaluation & Assessment Tracking (HEAT)
7. Pennsylvania State ATAC Program.
8. Violent Crime Linkage Analysis System (ViCLAS)

5.3 OPTICAL SATELLITE LINKS

5.3.1 Definition

*   **Free-space optical communication (FSO)** is an optical communication technology that uses light propagating in free space to wirelessly transmit data for telecommunications or computer networking.
*   In outer space, the communication range of free-space optical communication.

5.3.2 Block Diagram

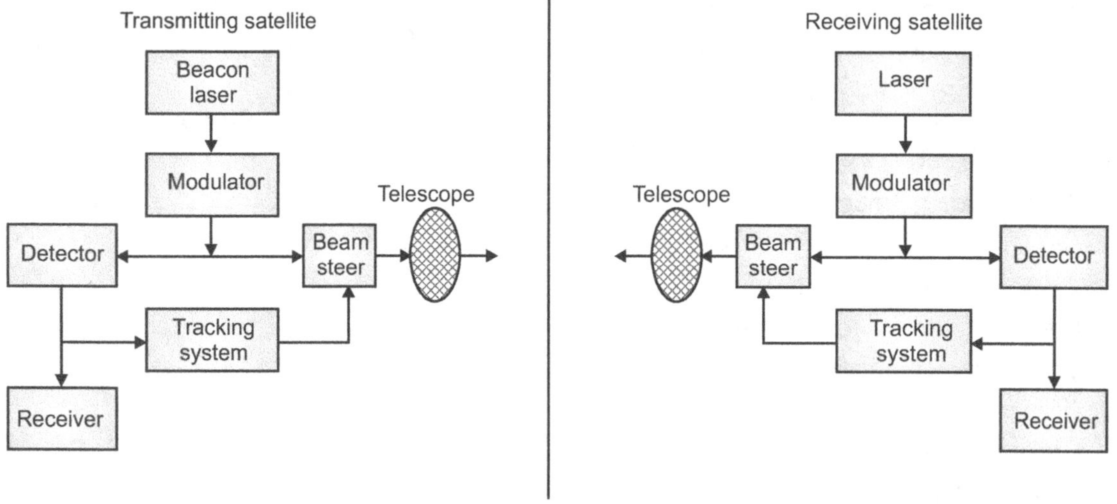

Fig. 5.3 : Block diagram of optical satellite crosslink

5.3.3 Laser Sources Used

*   Laser sources operate in single and multi-frequency modes. In single frequency mode, spectral width varies between 10 kHz and 10 MHz. In multi-frequency mode it is from 1.5 to 10 nm. The power emitted depends on the type of laser. Modulation can be internal or external. Internal modulation implies direct modification of the operation of the laser.
*   External modulation is a modification of the light beam after its emission by the laser. The intensity, the frequency, the phase and the polarisation can be modulated. Phase and polarisation modulation are external. Intensity and frequency modulation can be internal or external.
*   Polarization modulation requires the presence of two detectors in the receiver, one for each polarization. Because of this, it is preferable to reserve polarization for multiplexing of two channels.

5.3.4 Description

- The block diagram of optical satellite cross link model is shown in Fig. 5.2. In comparison with radio links, optical links have specific characteristics. Two aspects should be indicated:

- The small diameter of the telescope which is typically of the order 0.3 m. In this way one is freed from congestion problems and aperture blocking of other antennas in the payload.

- The narrowness of the optical beam which is typically 5 microradians. Notice that this width is several orders of magnitude less than that of a radio beam and this is an advantage for protection against interference between systems.

- But it is also a disadvantage since the beamwidth is much less than the precision of satellite altitude control (typically 0.10 or 1.75 mrad). Consequently an advanced pointing device is necessary; this is probably the most difficult technical problem.

- One of the major challenges in lasercomm is the development of a pointing, acquisition, and tracking (PAT) system to establish and maintain the laser links.

- There are three basic phases to optical communications:

  1. Acquisition
  2. Tracking
  3. Communications

5.3.5 Acquisition

- The beam must be as wide as possible in order to reduce the acquisition time. But this requires a high power laser transmitter. A laser of lower mean power can be used which emits pulses of high peak power with a low duty cycle.

- The beam scans the region of space where the receiver receives the signal, it enters a tracking phase and transmits in the direction of the received signal. On receiving the return signal from the receiver, the transmitter also enters the tracking phase. The typical duration of this phase is 10 seconds.

5.3.6 Tracking

- The beams are reduced to their nominal width. Laser transmission becomes continuous. In this phase which extends throughout the following, the pointing error control device must allow for movements of the platform and relative movement of the two satellites.

- In addition, since the relative velocity of the two satellites is not zero, a lead-ahead angle exist between the receiver line of sight and the transmitter line of sight. As will be demonstrated below, the Leadahead angle is larger than the beamwidth, and must be accurately determined.

5.3.7 Pointing

- Due to the narrow transmission (14.6 arc seconds) for the CLICK (Current Infrared Cross Link) terminal.

- PAT typically consists of a coarse pointing system with a wide field of view and limited precision supplemented by a fine pointing system with a narrow field of view but high precision.

- The CLICK coarse pointing system directly utilizes the spacecraft's attitude determination and control system (ADCS). The fine pointing system (FPS) uses a commercially available microelectromechanical system (MEMS) fast-steering mirror (FSM) controlled via feedback from a quadrant photodiode detector (quadcell), which also senses the beacon.

- In the payload optical system layout, there are three optical paths such as Beacon R_x signal and the communications T_x and R_x signals. The objective of the fine pointing system (FPS) is to align the communications T_x and R_x signals to within the pointing requirement.

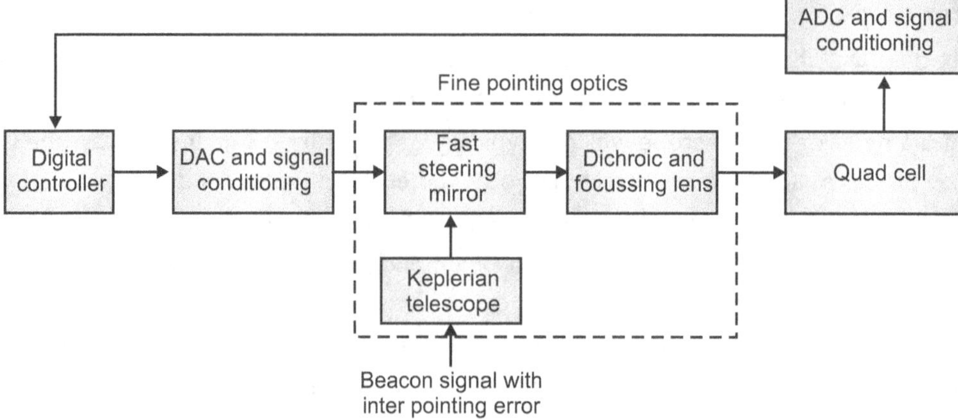

Fig. 5.4 : Fine Pointing System (FPS)

- The block diagram fine pointing system (FPS) is shown in Fig. 5.4. The component selection and optical design of the system is detailed in long like. The laser spot sensor is a First Sensor QP1-6 quadrant photodiode detector ("quadcell"), which consists of 4 Silicon PIN photodiode sensors evenly arranged in a rectangular pattern. The beacon signal is detected on the quadcell, and the output signals are amplified and filtered via a transimpedance amplifier and a bandpass filter, respectively.

- The analog output signals are sampled by an Analog to Digital Converter (ADC) and routed to a Raspberry Pi Compute Module 3, which uses a Broadcom BCM2837 processor. The X and Y centroids are calculated from the quadcell measurements and used by the controller.

- The resulting control signals are converted to analog signals, amplified, and then filtered via a 6th order Bessel low-pass filter before being routed to the Mirrorcle MEMS FSM. The FSM uses a Differential Quad-channel scheme to actuate a 3.6 mm diameter mirror, which has a maximum mechanical angle of ±3.2 degrees in both the X and Y axes.

5.4 CABLE CHANNEL FREQUENCY

5.4.1 Brief History

- In 1990, General Instrument (GI) demonstrated that it was possible to use digital compression to deliver *high quality* HDTV in a standard 6 MHz television channel. Using the same technology General Instrument (GI) demonstrated the digital transmission of multiple high quality standard definition programs in a 6 MHz cable channel.

- In the 1990s, cable providers began to invest heavily in this new multi-channel digital TV technology to expand the number of channels and services available to subscribers. Increased competition and programming choices from direct-broadcast_satellite services such as DirecTV, Dish_Network, and PrimeStar caused cable providers to seek new ways to provide more programming.

- Customers were increasingly interested in more channels, pay-per-view programming, digital music services, and high speed internet services. By 2000, most cable providers in the US were offering some form of digital cable TV to their customers.

- **Digital cable technology** has allowed cable providers to compress video channels so that they take up less bandwidth and to offer two-way communication capabilities. This has enabled providers to offer more channels, video-on-demand services that do not require a separate telephone line, telephone services, high speed internet services, and interactive television services.

- Digital cable implements error correction to ensure the integrity of the received signal and uses a secure digital distribution system (i.e. a secure encrypted signal to prevent eavesdropping and theft of service).

- Most digital cable providers use QAM for video services and DOCSIS standards for data services. Some providers have also begun to roll out video services using IPTV or Switched video.

5.4.2 Definition

- **Digital cable is the distribution of cable television using digital video compression for distribution.**
- The technology was developed by General Instrument, which was succeeded by Motorola and subsequently by ARRIS_Group.

5.4.3 Channel versus Frequency

- The cable TV channel frequencies mention channels 2, 3, 4, 5, 6, 7, 8, 9, 10, 11, 12, 13, channel A-2, A-1, A, B, C, D, E, F, G, H, I, J, K, L, M, N, O, P, Q, R, S, T, U, V, channel-W and Channel-AA to Channel-RR.
- As we know microwave frequencies cannot be transmitted over the cable. In order to transmit these high frequency signals over cable they are downconverted to lower frequencies using mixers. Few of these channels are mapped to standard 6MHz TV channel frequencies.
- The cable TV channel frequencies are nonbroadcast channels. Heterodyne processors are employed to translate received signals to the desired channels. The processor will have tuner which will pick the over the air signals and produce channels with IFs of about 45.75 and 41.25 MHz. These picture and sound IFs are separated by filters and passed through AGC circuits to provide gain.
- The table below mentions cable TV channels or cable TV frequencies.

Channel Number	Frequency, video carrier (MHz)	Channel Number	Frequency, video carrier (MHz)
Channel-2	55.25	Channel-M	235.25
Channel-3	61.25	Channel-N	241.25
Channel-4	67.25	Channel-O	247.25
Channel-5	77.25	Channel-P	253.25
Channel-6	83.25	Channel-Q	259.25
Channel-A-2	109.25	Channel-R	265.25
Channel-A-1	115.25	Channel-S	271.25
Channel-A	121.25	Channel-T	277.25
Channel-B	127.25	Channel-U	283.25
Channel-C	133.25	Channel-V	289.25
Channel-D	139.25	Channel-W	295.25
Channel-E	145.25	Channel-AA	301.25
Channel-F	151.25	Channel-BB	307.25
Channel-G	157.25	Channel-CC	313.25
Channel-H	163.25	Channel-DD	319.25
Channel-I	169.25	Channel-EE	325.25
Channel-7	175.25	Channel-FF	331.25
Channel-8	181.25	Channel-GG	337.25
Channel-9	187.25	Channel-HH	343.25
Channel-10	193.25	Channel-II	349.25
Channel-11	199.25	Channel-JJ	355.25
Channel-12	205.25	Channel-KK	361.25
Channel-13	211.25	Channel-LL	367.25
Channel-J	217.25	Channel-MM	373.25
Channel-K	223.25	Channel-NN	379.25
Channel-L	229.25	Channel-OO	385.25
Channel-PP	391.25		

5.5 HEAD END ELEVATION

5.5.1 Introduction

- Although mechanical ventilation is a life-saving measure, it is associated with respiratory complications such as atelectasis, ventilator-induced lung injury, ventilator-associated pneumonia, and ARDS. This results in increased mechanical ventilation, prolonged stays in the ICU and hospital, and increased costs to the health care system.

- Head-of-bed elevation (HOBE) is one of several therapeutic interventions that have been demonstrated to reduce respiratory complications associated with mechanical ventilation.

- Although literature is available on the effects of various positions on oxygenation, hemodynamics, and prevalence of pulmonary complications, there are few studies that have examined lung volume in relation to HOBE.

- Until recently, methods to monitor changes in lung ventilation in the ICU have been limited and challenging. For instance, the use of computed tomography imaging exposes patients to substantial doses of ionizing radiation and cannot be performed at bedside. This requires transporting patients out of the ICU, posing safety risks to patients, and incurring considerable costs. An alternative to computed tomography imaging is the relatively new imaging method of electrical impedance tomography, which is a noninvasive, radiation-free monitoring tool that allows real-time imaging of ventilation at bedside.

5.5.2 Need

- The height of the bed head elevation is critical and must be at least 6-8 inches (15-20 cm) to be at least minimally effective to prevent reflux of gastric contents. A meta-analysis of previous non-randomized studies revealed that bed head elevation is an effective therapy in relieving the symptoms of reflux.

- By sleeping elevated the mucus will be able to drain more easily and allow to sleep more peacefully. This position will also aid in preventing the mouth breathing brought on by respiratory problems. Lying flat can put too much pressure on the head causing the brain to get congested.

- Therefore, there is a need of head end elevation.

5.5.3 Sleeping Position

- By for the healthiest option for most people, sleeping on you book allow your head, neck and spine to rest in a neutral position. This means that there's no extra pressure on those areas, so you are less likely to experience pain. Sleeping facing the ceiling also ideal for warding off acid reflux.

5.5.4 Ideal Head Position

- **Neurointensivists have been greatly interested** in the **preferred degree** of **head elevation** in **patients** with **raised intracranial pressure (ICP). Elevating** the head and toss in one of the few easily instituted methods of reducing ICP; the ICP probably drops due to drainage of cerebral veneus blood and CSF. Although, this maneuver can reduce elevated ICP, reduces cerebral perfusion pressure (CPP).

- Blood pressure (BP), SPP and ICP all the head and upper body were elevated progressively from supine 0° to 15° and then to 30°. Blood pressure measured through a radial artery or femoral-artery cather and a transducer at the level of the foramen of Monro fell to a greater extent. Therefore, supine position was probably preferable with regard to perfusion pressure, but that obtained body position should be established individually and that in any case, routine use 0 to 30° elevation could not be endorsed.

5.5.5 Benefits

- By raising the head, one will open breathing passages to improve sleep problems due to health issues and snoring. Benefits of sleeping elevated allow those suffering from the illnesses such as those caused by heart diseases, acid reflux, and congestion problems, as well as sleep apnea.

- The benefits of sleeping elevated (or insured bed therapy) are us under:
    1. Stop Mouth Breathing
    2. End Post-Nasal Drop
    3. Curb Snoring and Sleep Apnea
    4. Alleviate Heart Congestion and Inflammation

5. Lessen Shortness of Breath
6. COPD Relief
7. Soothe Head Stuffiness
8. Relieve Migraine
9. Help for Nervous System, Spine and Movement Disorders

5.6 DISTRIBUTION OF SIGNAL

5.6.1 Definition

- In signal processing, a **signal** is a function that conveys information about a phenomenon in electronics and telecommunications.

- It refers to any time varying voltage, current or electromagnetic wave that carries information. A signal may also be defined as an observable change in a quality such as quantity.

- Signal analysis is a tool for achieving this aim the principle of signal analysis is to break up all signals into summations a sinusoidal components. This provides a description of a given signal in terms of sinusoidal frequencies.

- In information therapy, a signal is a confined message, that is, the sequence of states in a communication channel that encodes a message.

- In the context of signal processing, signals are analog and digital representations of analog physical quantities.

5.6.2 Common Distributions

- There are number of different random distribution of signals in existence, many of which map very well to natural phenomenon. The most basic and most common distribution of signals are as under:

  1. Uniform Distribution
  2. Gaussian Distribution
  3. Poisson Distribution

5.6.3 Uniform Distribution

- One of the most simple distributions is a Uniform Distribution. Uniform Distributions are also very easy to model on a computer, and then they can be converted to other distribution types by a series of transforms.

- A uniform distribution has a PDF that is a rectangle. This rectangle is centered about the mean, $<\mu_x$, has a width of A, and a height of 1/A. This definition ensures that the total area under the PDF is 1.

5.6.4 Gaussian Distribution

- The Gaussian (or normal) distribution is simultaneously one of the most common distributions, and also one of the most difficult distributions to work with. The problem with the Gaussian distribution is that its PDF equation is *non-integratable*, and therefore there is no way to find a general equation for the CDF (although some approximations are available), and there is little or no way to directly calculate certain probabilities.

- However, there are ways to approximate these probabilities from the Gaussian PDF, and many of the common results have been tabulated in table-format. The function that finds the area under a part of the Gaussian curve (and therefore the probability of an event under that portion of the curve) is known as the Q function, and the results are tabulated in a Q table.

5.6.5 Poisson Distribution

- The Poisson Distribution is different from the Gaussian and uniform distributions in that the Poisson Distribution only describes discrete data sets. For instance, if we wanted to model the number of telephone calls that are traveling through a given switch at one time, we cannot possibly count fractions of a phone call; phone calls come only in integer numbers.

- Also, you can't have a negative number of phone calls. It turns out that such situations can be easily modeled by a Poisson Distribution. Some general examples of Poisson Distribution random events are:
  1. The telephone calls arriving at a switch
  2. The internet data packets traveling through a given network
  3. The number of cars traveling through a given intersection

5.6.6 Probability Distribution

- A *probability distribution* **p(x)** may be defined as a non-negative real function of all possible outcomes of some random event. The sum of the probabilities of all possible outcomes is defined as 1, and probabilities can never be negative.

- For example, a *coin toss* has two outcomes, "heads" (H) or "tails" (T), which are equally likely if the coin is "fair". In this case, the probability distribution is

$$p(H) \ = \ p(T) = \frac{1}{2} \qquad \qquad ...(5.1)$$

where **p(x)** denotes the *probability* of outcome **x** . That is, the total ``probability mass'' is divided equally between the two possible outcomes heads or tails. This is an example of a *discrete* probability distribution because all probability is assigned to two discrete points, as opposed to some continuum of possibilities.

5.6.7 Statistical Distribution

- Inequality measurement in its modern form derives some of its intellectual heritage from statistics and is considered by some to be a branch of welfare economics. The subject is related to, but distinct from, the measurement of concentration, poverty, or relative deprivation. Inequality measurement in economics is perhaps most often applied to the study of income distributions but is also used in connection with the study of wealth, expenditure, and other indicators of well-being.

- Inequality measurement is principally concerned with the comparison of statistical distributions of a specific economic or social indicator. The comparisons usually, but not necessarily, involve distributions among persons or households and involve quantitative methods. The motivation for considering inequality measurement in this way is both analytical and practical.

- Outlier detection using statistical methods outlier in the data can be identified by creating a statistical distribution model of data and identifying the data points that do not fit into the model or data points that occurring the end of the distribution tails.

5.7 NETWORK SPECIFICATIONS

5.7.1 Definition

- A **network** is a collection of computers, servers, mainframes, network devices, peripherals, or other devices connected to one another to allow the sharing of data.

- An excellent example of a network is the Internet, which connects millions of people all over the world. To the right is an example image of a home network with multiple computers and other **network devices** all connected.

5.7.2 Network Devices

1. Desktop computers	2. Laptops
3. Mainframes	4. Servers
5. Consoles and thin clients	6. Firewalls
7. Bridges	8. Repeaters
9. Network Interface cards	10. Switches
11. Hubs	12. Modems

13. Routers	14. Smartphones
15. Tablets	16. Webcams

5.7.3 Network Topologies

1. Bus topology	2. Mesh topology
3. Ring topology	4. Star topology
5. Tree topology	6. Hybrid topology

5.7.4 Physical Networks

- The basic design for the IBM Pure Data System for operational Analytics uses three physical network as under :

  1. Internal application network

  2. Corporate network

  3. IBM Hardware Management Console (HMC) network

- The network infrastructure is supported by the following hardware :

  1. 1 Gbps Ethernet fabric

  2. 10 Gbps Ethernet fabric

- The network configuration for the IBM pure Data System for Optional Analytics is automatically configured, when the system is deployed

5.7.5 Allowed Traffic on Platform

- Only one MAC address per member is allowed. This limits the risk of loop on the network. Only 3 Ether types are allowed. They are as under :

  1. 0x0800 - IPv4

  2. 0x0806 - ARP

  3. 0x86dd - IPv6

- ICMPv6 and ARP traffic are "rate-limited"-"unknown-unicast" traffic, which is broadcasted by the platform is also "rate limited".

- A global filter is denying STP and bridging protocols. Link layer protocols and IPv6 Router Advertisement/Router Solicitation (RA/RS) are also filtered by the platform.

5.7.6 Network Equipment Specifications

1. Area Network Type	2. Form Factor
3. Network Media	4. Number of Ports
5. Data Rate	6. Packet Forwarding Rate

5.7.7 Network Types

1. Personal Area Network (PAN)	2. Local Area Network (LAN)
3. Wireless Local Area Network (WLAN)	4. Campus Area Network (CAN)
5. Metropolitan Area Network (MAN)	6. Wide Area Network (WAN)
7. Storage-Area Network (SAN)	8. System-Area Network (SAN)
9. Passive Optical Local Area Network (POLAN)	10. Enterprise Private Network (EPN)
11. Virtual Private Network (VPN)	

5.7.8 Network Software Specifications

- In order to guarantee the security of the exchange point, a set of rules has been defined. France-IX reserves the right to shutdown ports that violate these specifications :

  1. The MTU size should be 1500 bytes

  2. Non-unicast packets are not allowed except:

      (i) ICMPv6 Neighbor Advertisement/Solicitation

      (ii) ARP

      (iii) IPv4 multicast is not allowed

- To ensure that these rules are observed:

  1. A quarantine VLAN is used before moving a port into production to check that these specifications are followed

  2. A monitoring tool alerts the technical team if a new ARP entry is detected. Proxy ARP configured on the member port is not allowed.

  3. Sniffer servers are installed in the core network to analyse broadcast traffic, and check that only legitimate traffic is forwarded on the platform.

5.8 OPTICAL FIBRE CATV SYSTEM

5.8.1 Brief History

- A CATV or cable TV began in the United States as a commercial business in 1950, although there were small-scale systems by hobbyists in the 1940s. The early systems simply received weak (broadcast) channels, amplified them, and sent them over unshielded wires to the subscribers, limited to a community or to adjacent communities.

- The receiving antenna would be higher than any individual subscriber could afford, thus bringing in stronger signals; in hilly or mountainous terrain it would be placed at a high elevation.

- At the outset, cable systems only served smaller communities without television stations of their own, and which could not easily receive signals from stations in cities because of distance or hilly terrain.

- Although early (VHF) television receivers could receive 12 channels (2-13), the maximum number of channels that could be broadcast in one city was 7: channels 2, 4, either 5 or 6, 7, 9, 11 and 13, as receivers at the time were unable to receive strong (local) signals on adjacent channels without distortion. (There were frequency gaps between 4 and 5, and between 6 and 7, which allowed both to be used in the same city).

- As equipment improved, all twelve channels could be utilized, except where a local VHF television station broadcast. The bandwidth of the amplifiers also was limited, meaning frequencies over 250 MHz were difficult to transmit to distant portions of the coaxial network, and UHF channels could not be used at all.

- To expand beyond 12 channels, non-standard "midband" channels had to be used, located between the FM band and Channel 7, or "superband" beyond Channel 13 upto about 300 MHz; these channels initially were only accessible using separate tuner boxes that sent the chosen channel into the TV set on Channel 2, 3 or 4.

- Once consumer TV sets had the ability to receive all 181 FCC allocated channels, premium broadcasters were left with no choice but to scramble. Later, the cable operators began to carry FM radio stations, and encouraged subscribers to connect their FM stereo sets to cable. Before stereo and bilingual TV sound became common, Pay-TV channel sound was added to the FM stereo cable line-ups.

- During the 1980s, United States regulations not unlike public, educational, and government access (PEG) created the beginning of cable-originated live television programming. As cable penetration increased, numerous cable-only TV stations were launched, many with their own news bureaus that could provide more immediate and more localized content than that provided by the nearest network newscast.

- Although for a time in the 1990s, television receivers and VCRs were equipped to receive the mid-band and super-band channels. The conversion to digital broadcasting has put all signals – broadcast and cable – into digital form, rendering analog cable television service mostly obsolete,

5.8.2 Definition

- CATV originally stood for Community Access Television or Community Antenna Television. The abbreviation CATV is often used for cable Television and it is originated in 1948.

- **Cable television** is a system of delivering television programming to consumers via radio frequency (RF) signals transmitted through coaxial cables, or in more recent systems, light pulses through fiber-optic cables.

5.8.3 Distribution

- To receive cable television at a given location, cable distribution lines must be available on the local utility poles or underground utility lines. Coaxial cable brings the signal to the customer's building through a *service drop*, an overhead or underground cable.

- If the subscriber's building does not have a cable service drop, the cable company will install one. The standard cable used in the U.S. is RG-6, which has a 75 ohm impedance, and connects with a type F connector. The cable company's portion of the wiring usually ends at a distribution box on the building exterior, and built-in cable wiring in the walls usually distributes the signal to jacks in different rooms to which televisions are connected.

- Multiple cables to different rooms are split off the incoming cable with a small device called a splitter. There are two standards for cable television; older analog cable, and newer digital cable which can carry data signals used by digital television receivers such as HDTV equipment. All cable companies in the United States have switched to or are in the course of switching to digital cable television since it was first introduced in the late 1990s.

- Most cable companies require a set-top box or a slot on one's TV set for conditional access module cards to view their cable channels, even on newer televisions with digital cable QAM tuners.

5.8.4 Principle of Operation

- In the most common system, multiple television channels (as many as 500, although this varies depending on the provider's available channel capacity) are distributed to subscriber residences through a coaxial cable, which comes from a trunkline supported on utility poles originating at the cable company's local distribution facility, called the "headend". Many channels can be transmitted through one coaxial cable by a technique called frequency division multiplexing.

- At the local headend, the feed signals from the individual television channels are received by dish antennas from communication satellites.

Practice Questions

1. What is Laser ?
2. Explain principle of working of Laser.
3. List applications of Laser.
4. Describe Link analysis satellite communication.
5. Describe optical satellite link with its block diagram.
6. Describe satellite beam acquisition, tracking and pointing.
7. What is cable channel frequency ?
8. What is head-end elevation ?
9. What are benefits of head-end elevation ?
10. Why do you need head elevation ?
11. What is distribution of signals ?
12. Describe uniform, gaussian, and pointing distribution of signals.
13. State network topologies and devices.
14. State specifications of network.
15. Describe optical fiber CATV system.